FIRE IN
PARADISE

FIRE IN
PARADISE

AN AMERICAN TRAGEDY

ALASTAIR GEE AND DANI ANGUIANO

W. W. NORTON & COMPANY
Independent Publishers Since 1923

For information about permission to reproduce selections from this book, write to
Permissions, W. W. Norton & Company, Inc., 500 Fifth Avenue, New York, NY 10110

For information about special discounts for bulk purchases, please contact
W. W. Norton Special Sales at specialsales@wwnorton.com or 800-233-4830

Manufacturing by Lake Book Manufacturing
Book design by Chris Welch
Production manager: Lauren Abbate

ISBN 978-1-324-00514-8

W. W. Norton & Company, Inc., 500 Fifth Avenue, New York, N.Y. 10110
www.wwnorton.com

W. W. Norton & Company Ltd., 15 Carlisle Street, London W1D 3BS

1 2 3 4 5 6 7 8 9 0

Leslie Big Bird,   7-8-2020
    Howdy to you in enchanted land!
Here's a very disturbing read but
must be told. It's breaks my
heart but I had to know the
truth & facts as to why. If
you want to cut to chase ASAP
turn to last chp 'Perfect fire'
p-202-235 for wid ence on (PG+E's)
negligence, greed + apathy. Also
POTUS Chump's bully arragant respone
on p.173-4. Tired of being pushed
around-time for justice +fair
$settlement in mid-Aug-Sept. one
more leg surgery knee r plantin
late Sept

                        Stay safe-
Enjoyed "Zoom" event hog X I
couldn't see anyone   BEST,
            Burnt Bear on the MEND!
              Rollin' Schoeler

CONTENTS

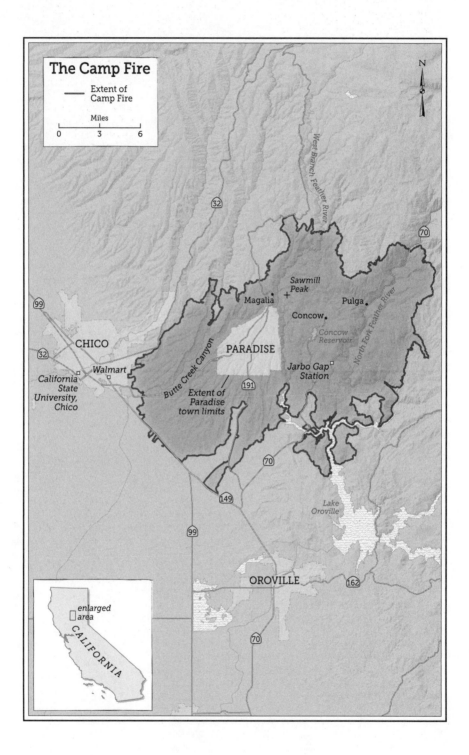

The Camp Fire

Extent of
Camp Fire

Miles
0   3   6

N

32

West Branch Feather River

70

Sawmill
Peak
Magalia   Pulga
Concow
Concow
Reservoir

North Fork Feather River

99

CHICO

32
Walmart

California
State
University,
Chico

Butte Creek Canyon

PARADISE

Jarbo Gap
Station

Extent of
Paradise
town limits

191

70

149

99

Lake
Oroville

OROVILLE   162

70

enlarged
area

CALIFORNIA

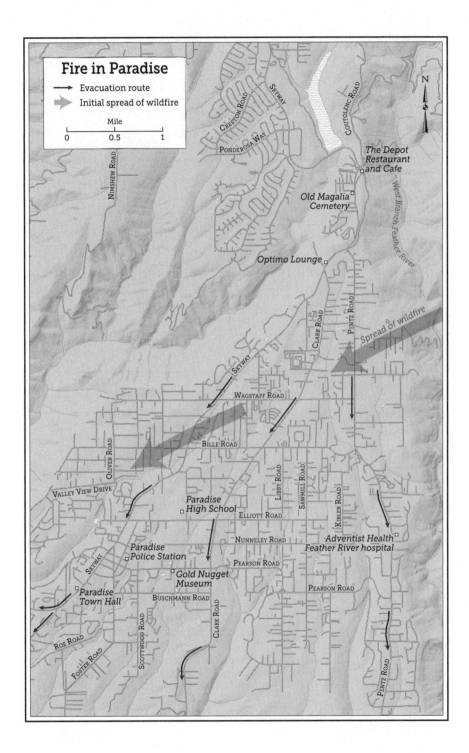

Prologue

At 8:30 a.m. on November 8, 2018, John Sedwick pounded on his daughter's bedroom door. "Fire's coming up over the ridge, Skye," he told her, his voice calm but loud, a sign of his increasing deafness. Skye got up quickly.

Sedwick was not an alarmist. He knew fire, and if he considered it serious, it was. As part of a career that included periods as a museum docent, logger for the forest service, and orderly at a state hospital, Sedwick had spent years as a volunteer firefighter on the Paradise Ridge, a forested perch in the foothills of California's Sierra Nevada mountain range. He'd been living here in the community of Magalia, just north of Paradise, off and on since he first visited as a child. More than seventy years later, Sedwick still resided in the same home, a century-old two-bedroom cabin with orange walls and a bathtub on the back porch. Skye, who had recently turned fifty, had moved in with him a couple years ago, when she was recovering from various health problems.

As smoke lingered in the air, Skye, fresh out of the shower, hurriedly gathered the things she'd need: clothes, shoes, and the makeup bag she never left home without.

The Sedwicks' rambling property was dotted with John's vintage vehicles, including an ancient flatbed truck and a faded red 1940s-era bulldozer its environs were magnificent. The inhabitants of the town of Paradise, and the smaller communities that surrounded it, spent their days among a gorgeous profusion of pines and oaks and wildflowers. More and more it was becoming a refuge for people priced out of the rest of the state, where the cost of living had risen to untenable levels.

Occasional wildfires broke out near the Ridge. But Sedwick, and nearly anyone who had ever paid much attention, knew that the area could face a real fire, a bad one. People prepared for the worst, but they didn't always comprehend how awful the worst could be. When the fire risk was high, Sedwick always made the same plea as he and his daughter sat down for meals: "Please Jesus, no fires today."

After packing, Skye walked outside to where her father stood in the yard. "Dad, is there something I should do? Should I wet the ground?" Sedwick shook his head. When she returned with her bags, Sedwick told her that he would stay, for now. Skye, an auburn-haired artist who used to work as a counselor in a juvenile detention center, was planning to meet her boyfriend at a gas station a mile or so down the road. She took a few steps and paused. Still holding her bags and her dog, Skye hugged her trim, compact father goodbye. He grunted when she said *I love you*, but she knew that meant *I love you, too*.

The road that curved past their home was already blocked and nearly at a standstill with cars trying to evacuate. Don't go that way, Sedwick had warned Skye. She set off down a quiet lane leading away from their house, shaded by ponderosa pines and Douglas firs, until she was just far enough from the cabin that Sedwick wouldn't have heard her if she yelled. She could see flames catching in the undergrowth, and thought about turning back to tell him how bad things were but decided against it. *He knows*, she thought to herself. Sedwick was stubborn. There was no getting him to change tack once he had made a decision. *Dad will leave if it gets bad.*

She moved even more quickly, through the smoke and up onto an embankment that she used where the road became impassable. Ethereal bits of glowing dust fell from the sky. Skye watched them uneasily. *Embers*, she thought.

It was approximately 9:30 a.m. A few hours earlier, satellites had captured flames approaching Paradise at about 21 miles per hour, faster than the previously posited maximum for wildfire spread. They were consuming almost four hundred American football fields' worth of vegetation a minute.

Smoke blotted out the sun and streamed low over the town, reducing the visible world to Paradise and nothing more. The flames, when they arrived, were unstoppable, far beyond the capacity of any firefighter to control. In places they burned as hot as a crematorium. Cars turned into rivulets of molten metal. Homes became matches that set fire to the next. The conditions gave rise to apparitions only seen in the most extreme of blazes: fire tornadoes.

Within hours, the fire would consume Paradise, trapping thousands of people who were trying to flee on roads designed for a fraction of that number. Emergency responders would mount a heroic defense and rescue operation that included at its peak 5,500 firefighters and 600 engines, but when the blaze was finally brought under complete control with the help of rain seventeen days later, it had become the deadliest wildfire in America in a century, and the deadliest ever in recorded California history. The town of Paradise would suffer a fate almost without parallel in the modern era in the United States: total devastation.

PART ONE
PARADISE

1

A Gold Rush Town

John Sedwick and his daughter Skye had spent the summer not talking to each other, an impressive feat considering they lived together. It had started, as arguments often do, with something banal: a bathroom lamp broke. Sedwick blamed Skye. Nobody does the silent treatment like a Scorpio, Skye, a Libra, would think.

But finally, after a long, hot summer on the Paradise Ridge, and three months of separate meals, Skye had written a note to her father: *Heaven forbid something happens to either of us when we're not speaking to each other. You and I both know life is short.*

So on November 4, 2018, here they were at the Depot restaurant, a quaint 115-year-old former train stop just north of Paradise, for a birthday dinner marking Sedwick's own eighty-two years on the planet. Money was hard to come by for the both of them, but Skye had sold one of her bricolage paintings and decided this was the best thing on which she could splurge.

More than a business, the Depot was the heart of Magalia. It was built in 1903 as a stop on a thirty-one-mile train line that connected the valley with a lumber mill in the mountains. Since the railroad closed, the Depot had new lives as a flea market and various restaurants. It was also a welcome sign of sorts, reminding residents of this area that they were home, and Skye knew how much it meant to her father. His passion for the Depot was matched only by his fondness for exploring and excavating the history of the Ridge.

Sedwick had spent his early years in the Bay Area, though each summer his parents, an Oakland police clerk and his Scottish immigrant wife, who worked at an upscale San Francisco department store, took Sedwick and his two older sisters up to the community of Magalia. The country air helped his asthma, and his family moved permanently when he was around ten. He still lived in the same home, a cabin that had been built sometime around 1906. It was so venerable, at least for this part of the country, that the family couldn't find a proper record of its construction with the county assessor.

With its questionable electrical system and single-wall construction, the home wasn't really up to modern standards, but Sedwick wanted to keep it as it had been when his parents were alive, which his family suspected was a way for him to maintain his close connection to his mother and the other family members who had lived there over the decades. A tender portrait of Sedwick's mother at age twenty hung by the entryway. Sedwick had been close to her, a woman ahead of her time who believed in open marriage and wrote a letter to the *San Francisco Chronicle* protesting the internment of Japanese-Americans during World War II. He named Skye for the Scottish island his mother's mother was born on.

There was no lock on the door, just a piece of wood Sedwick and Skye used to secure it at night. In the living room, next to Sedwick's record collection, was a woodstove that had been in the cabin for more than a hundred years. They used it every day in the winter—it was the cabin's primary source of heat. Sedwick went to bed early, and Skye, an insomniac, would keep an eye on it through the night. The year before, as Skye was washing her paintbrushes late one December evening, she saw a chimney fire had started. She roused her father, who put it out and went back to bed. Yet fire was soon creeping along the walls adjacent to the stove. She woke Sedwick again. Always level-headed and quick to act, he climbed into the rafters of the house to tamp it out. "You're a badass, Dad," Skye told him the next morning. In that moment, she'd gotten a glimpse of what a fine firefighter he had been.

Over the decades, Sedwick preserved his historic home and relished traditions like the annual Gold Nugget Days, when thousands thronged a Gold Rush–themed parade, a donkey race, and a fair. It had run for almost sixty years, and commemorated the mad scramble for gold in the very hills on which Paradise stood, seen from a nostalgic perspective as a time of self-sufficiency and quaint dress.

Meanwhile the Ridge changed around Sedwick. He watched it morph from isolated settlements tucked in the trees to a bigger, louder place. The Paradise High School graduating class increased from around 85 in 1957, according to the yearbook, to 220 in 2018. Locals recalled that the first stoplight was installed on the main thoroughfare in the late 1960s, and in recent years so much traffic roared down it that in some parts the town narrowed the road to two lanes. It was partly to reduce accidents and partly to encourage people to stop at the downtown businesses.

Sedwick was Old Ridge, but Skye had struggled with New Ridge problems. Meth, opioids, and, of late, heroin were taking a toll. In the early 1990s, Skye, battling addiction, fled to the East Coast where she got sober, raised two kids, and earned a master's degree. When she moved back to the Ridge from Albuquerque in 2016, she was recovering from a flare-up of Graves' disease, an autoimmune disorder that causes an overactive thyroid and, in turn, rapid weight loss and muscle weakness. Skye and her dad, who was lonely after the death of his third wife and great love and was close to retiring from an industrial supplies store in Chico, had reached a decision together: she would come back and focus on her health, and she'd also help out with the cooking and cleaning.

Despite their occasional spats, and the distance from her adult daughter and son, who lived in New Mexico, Skye was happy in Magalia with her quirky father, who greeted loved ones with "Yo" and sometimes wore a My Little Pony T-shirt. He'd been working for years on an espionage thriller that took place on a train traveling through a nearby canyon—he hadn't finished it yet, but he could recite entire chapters from memory. Shortly after she arrived, Sedwick got her a dog for her birthday, a brown-eyed mutt from the shelter she named Jude. He took to Skye right away, occasionally breaking out of the yard to find her when she would leave to run errands. Skye loved the mornings she woke up to her dad playing the steel pedal guitar and singing like Willie Nelson, and the time they spent picking fruit from the peach and plum trees planted by his mother. And they were both great walkers.

One of their favorite strolls was along a trail where railroad tracks once ran by the Depot. It is a gradual climb uphill, past gray pines and Indian paintbrush. Along the way, Sedwick would admire the serpentine, a subtle, greenish rock whose presence in the Sierra is associated with distinctive plant communities and with gold, while Skye gathered iris and poppies that she would press and use in collages. The path comes to an end by a pond, near a dip in the earth where the remnants of an old cabin are visible. This stroll captured everything Sedwick loved about the Ridge: the trees, the railroad, and glimpses of the Feather River

Canyon, parts of which, he once wrote, "will make a believer out of anyone who doubts God's handiwork." He felt at home anywhere on this whole plateau. The stories he wrote for a local historical publication ended with this biography: "John Sedwick grew up on the Paradise Ridge; its hills, forests, streams, people, buildings and trains are part of his life."

This part of the state, 150 miles northeast of San Francisco, in the foothills of the Sierra Nevada, felt a world apart from the clichéd California of beaches and wineries, the billionaires of Silicon Valley and Hollywood. It wasn't an international tourist destination and it mostly wasn't on the road to anywhere else. The town had really made it, a local joke went, when a Starbucks opened in spring 2018. Yet in another way, Paradise, cocooned in nature, was Californian to the core.

The symbol on the California flag is the grizzly bear, but it might as well be the redwood, the palm, or the Joshua tree, for all that these evocative plants populate and frame our imaginings of the Golden State. Early settlers described it as a Pacific Eden and "the garden of the world." California means Yosemite, Sequoia, and Muir Woods; it is almond and avocado trees, grapevines and strawberry fields. At the core of the California Dream, the fantasy of California life as it emerged in the formative years of the state's existence, was "the hope for a special relationship to nature," historian Kevin Starr has suggested. In one of those rare instances, the fantasies are justified.

California is an ecological island, cut off in the west by the Pacific Ocean and in the east by the forbidding deserts of the southwestern US, and its plant life has evolved in isolation in spectacular fashion. There are more than 2,000 endemic plants in California, outnumbering any place of similar size in North America. It is home to both the largest tree and the tallest tree in the world, redwoods that are so distinguished as to have been named. Hyperion, as one is called, soars to an unmatched 380 feet, while General Sherman is over 100 feet in circumference at its base, and one of its branches was recorded as taller than most trees east of the Mississippi. The General is a little over 2,000 years old, yet this seems fleeting compared to world's oldest individual tree, Methuselah, a 5,000-year-old bristlecone pine in California's White Mountains, which sprouted at the same time as the civilization of ancient Egypt. Bristlecones grow with as little as 12 inches of precipitation a year—a desert is defined as receiving less than 10 inches—and at such an imperceptible pace that their rings must be counted with a microscope. Sometimes comprised more of dead wood than living, with only the occasional nimbus of needles, their twisted and animal-striped branches stretch imploringly toward the sky.

In the desert, remarkable vegetation is invisible in plain sight. The creosote bush, a humble and omnipresent plant, produces clones of itself that grow outwards from the original plant and form a ring. The oldest, called the King Clone, has an average diameter of 45 feet and started life as much as 11,700 years ago. As for California's more evanescent flora, its springtime blooms of wildflowers—poppies in outrageous orange, lupines of pure violet, goldfields the color of the sun—were described by Henry James, in a 1905 letter to his sister-in-law, as "worthy of some purer planet than this."

Since California gained statehood in 1850, its inhabitants have supplemented nature's bounty. Or, in the old-fashioned term used by historian Jared Farmer, they have "emparadised" California. They have cultivated non-native plants by the million: eucalyptus trees from Australia, citrus from Europe, skyscraping Mexican fan palms and statement-making Canary Island date palms. And they have added all the luxuriant foliage and indulgent blooms a nature-minded Californian could want, from South American jacaranda trees, haloed with purple blossom, and bougainvilleas, wall climbers of saturated crimson and pink, to the South African bird of paradise, whose inflorescences appear to be taking flight on neon wings. In the unending sunshine and fertile soils in the center of state, meanwhile, much of America's fresh food is grown—over a third of its vegetables, and two-thirds of its fruits and nuts.

Lured by greenery, and the prospect of an outdoorsy life, millions of people have moved out to forested or wilderness areas, such as the rolling chaparral around Los Angeles and San Diego, or the foothills of the Sierra Nevada, the backbone of the state stretching 400 miles north to south. Fully one-third of homes in California are surrounded by or adjacent to wildlands. And yet these idyllic locations have another face.

California has a Mediterranean climate, like only four other regions around the world—the Mediterranean basin itself, the central coast of Chile, the Cape region of South Africa, and southern and southwestern Australia. In these areas it rains plentifully in the winter but far less, if at all, in the summer. Part of the reason California's plants are so diverse is because they have evolved to cope with supreme deprivation. By October, or November if the rains are late, the California countryside is desiccated, all verdure baked out of it. It is as dry as tinder.

Because fire is generally regarded as a hostile force, a marauder that must be defeated, it can be difficult to grasp that in California, fire is as natural and necessary as the rain or wind. Giant sequoias, like some other conifers, require fire to propagate themselves. Thick bark on mature trees enables them to survive a flame front, yet the ascending heat opens cones hanging on branches. Falling to earth, the seeds are able to germinate in sunlight because the fire has cleared away the leaf litter. Certain plants, like golden yarrow, whispering bells, and the fire poppy, are known as "fire followers," whose seeds are stimulated to germinate by the presence of compounds in burned wood. In desert oases, California fan palms rely on fire to kill other plants competing for precious moisture. Plants even appear to encourage fire. In the presence of flames, chamise, the most widely distributed plant in California's arid grasslands, emits combustible gases from its leaf surfaces.

Fire molded the very form of the landscape, by killing off young trees and dense shrubbery. It created open, sunny forests through which you could ride a horse, it was once said, without being snagged by branches on either side. Frequent lower-intensity fires—running through grassland, licking the bases of trees but not climbing into their crowns, perhaps smoldering for months in the duff—also reduced the risk of apocalyptic blazes that reduced entire wooded hillsides to char, because flammable materials were prevented from building up.

This was apparent to California's Native Americans, of whom there were 310,000, speaking a hundred languages, prior to European colonization. One romanticization of Native peoples holds that they left barely a footprint on the land, but in fact they managed California extensively—as ecologist M. Kat Anderson put it, they "tended the wild." This included pruning trees and shrubs, irrigating and watering the land, and, most important, setting fires to prompt new growth of preferred plants, to clear the terrain and make hunting easier, and to lessen the threat of big fires. Before Europeans came to California in the sixteenth century, as much as 4.5 million of California's 105 million acres burned every year, sparked by lightning or deliberately by people. It was, in the words of other researchers, "a gigantic hearth." Or, seen through contemporary eyes, a heaven fated for immolation.

"You are ascending into Paradise," read the sign by the roadside. "May you find Paradise to be all its name implies," said a second.

Just the journey there could put you in a more rarefied frame
of mind.

Paradise is a twenty-minute drive from the vast plains of California's agricultural Central Valley, on a road fittingly named
Skyway, which leads into the Sierra Nevada foothills. Up and up,
drivers suddenly crest a ridge and are confronted with plunging,
tree-speckled gorges that encircle the town and wind their way
into the inscrutable distance. The everyday affairs of the valley
floor seem to diminish in significance. Quickly the road enters
the tree line, under a canopy of soaring ponderosa pines and gra-
cious black oaks that enchanted many of its 27,000 inhabitants.
Another 15,000 or so lived in the unincorporated communities
that sprawled off up the hills or down into the valleys surrounding Paradise: Magalia, Butte Creek Canyon, Concow, and others.

Through the decades that California was under Spanish and
then Mexican control, this part of the state attracted little attention
from outsiders. Everything changed on January 24, 1848, when
a carpenter named James Marshall went to inspect a construction project he was supervising: a water-powered lumber mill on
the American River, northeast of present-day Sacramento. In the
channel through which water was to flow and drive the wheel,
he saw a fateful glint. A few months later, a settler ran into San
Francisco with news of the discovery of gold, and soon President James K. Polk announced it to the nation. California's non-
Native population soared from about 10,000 to almost 100,000
by the end of the next year as seekers of riches from around the
world descended, despite the fact that it could take five to eight
months to reach the state from the eastern US overland or by
boat. A San Francisco newspaper reported: "The whole country, from San Francisco to Los Angeles and from the sea shore
to the base of the Sierra Nevada, resounds with the sordid cry of
'gold! GOLD!! GOLD!!!' while the field is left half planted, the
house half built, and everything neglected but the manufacture of

shovels and pickaxes." The paper itself ceased publication because the town had emptied, with six hundred abandoned ships in the harbor. Those heading for the goldfields called themselves the Argonauts, after the band that accompanied Jason on his quest for the Golden Fleece.

Their hopes to hit pay dirt focused on the Mother Lode, a 120-mile stretch of the central Sierra Nevada. First they panned in riverbeds, and when these had been exhausted, they broke hillsides apart and felled thousands of trees. One contemporary compared it to cutting off a princess's fingers to obtain her jewels. Having voyaged into mountains lacking any infrastructure, theirs was a brutally hard existence inflected with sickness, murder, and lynch law. "For a few brief years in far-off California," wrote Starr, the historian, Americans "returned en masse to primitive and brutal conditions, to a Homeric world of journeys, shipwreck, labor, treasure, killing and chieftainship." Despite the fact that $2 billion worth of gold was recovered in total, many Argonauts found they could barely make ends meet. Instead of gold mining they assumed more prosaic occupations: peddler, shop assistant, farmer.

The largest pure nugget ever found—as opposed to an agglomeration of gold and another mineral—weighed 54 pounds and was called the Dogtown Nugget for a haphazard settlement to the north of the Mother Lode that was later renamed Magalia. It sold for $10,690. The man who found it in 1859 didn't come to much; he built a grand, colonnaded hotel in the valley that burned down twice, and he returned penniless to the goldfields. But the ridge enticed settlers, like Hirsch Cohn, a Jewish emigrant from Poland, who ran an emporium where gold dust could be exchanged for dynamite, or James Dresser from England by way of Canada, Ohio, and Illinois, who wrote to his daughters back east: "The climate is the nearest perfection of any we know anything about." These settlers were the nucleus of the communities from which Paradise, located on a triangular ridge between the West Branch Feather River and Butte Creek canyons, would soon emerge.

As in the rest of California, however, the prospectors were actually intruders. The Ridge and surrounding areas were already home to the Konkow, part of a tribal group of some 9,000 people. They camped at higher elevations in the summer and descended into the valley in the winter. In 1833 they had been decimated by an epidemic of what was probably malaria, introduced by foreign fur trappers. Amid the Gold Rush, their fate was sealed. "What is lower in the scale of humanity than a California Indian?" the *Californian* asked. "Their destiny is to be exterminated," said one gold merchant.

Suddenly they became outcasts in their own homeland. They starved because white men hunted the game, and they were killed "with no more compunction than if they were killing a coyote," one Konkow man said. A letter from one white settler to his relatives relates as much: "Wee killed one Indian 2 squas 2 children. . . . Then we took our breakfast which was crackers and Bacon and not enough of that."

The US negotiated a treaty with the Konkow that would have given them a swathe of territory in the area, but it was never ratified by senators, and survivors were eventually marched to a reservation over 100 miles away, where a Konkow leader reported that his people starved. That journey, in which at least 32 of the 461 unwilling marchers died along the way, and only 277 reached their destination, has become known as the Konkow Trail of Tears. By 1910, there were less than 1,000 Konkow remaining, mirroring the precipitous decline of the California Indians as a whole, whose population plummeted to 30,000 after the Gold Rush, or a tenth of what it had been before contact. As a result, the Paradise Ridge was left to the settlers.

The government received an application to create the Paradise post office, which listed a population of eighty, and it opened in 1877. It is rumored that a saloon called Pair O'Dice was the source of the name; more likely, it came from an appreciation of the area's natural qualities. Gold did not make the area rich, and it even earned the nickname of Poverty Ridge. Yet as one early settler proclaimed, "It's a darn nice place to starve if you have to." Residents turned to the cultivation of apples and olives and to the lumber industry. In 1923, a brochure proclaimed, "WE WANT MORE PEOPLE," and boasted "Abundant pure, soft, healthful well water. No malaria, hay fever, asthma, fleas or pesty insects. . . . American community with modern city conveniences."

It could be cozy or claustrophobic, depending on your out-look. Paradise just after World War II, for instance, was the kind of place where the local newspaper published the social comings and goings of residents: "Mrs. Frances Easley seen leaving for San Francisco to visit her sick mother," a "lingerie shower" for the wedding of Mrs. Sadie McFall: "The home was decorated with manzanita blooms and violets. . . . Refreshments of cake with whipped cream . . . and coffee were served." A resident wrote a column documenting her travels out of state: "When I tell people that I'm from Paradise, they look in amazement at me so earthly."

Like other rural, inland parts of California, Paradise was, and remained, conservative and religious, and in 1963 it gained national fame in *LIFE* magazine as a redoubt of the ultra-right-wing John Birch Society. The scandal concerned a high school social-studies teacher named Virginia Franklin, who encouraged vigorous political debate in her classes, organizing mock conventions and trips to an annual human rights conference. But the discovery that the conference was endorsed by the American Civil Liberties Union led to accusations from locals that Franklin was a

Communist, and after she gave her students a reading assignment from *The Nation* magazine, critics found that the article appeared "next to an advertisement for a book on conjugal techniques" and condemned her for promoting salaciousness. With the help of his father, one student smuggled into class a tape recorder placed in a hollowed-out book, hoping to capture incriminating evidence. The effort failed, as did an attempt to stack the school board with Franklin's opponents. Hearing the board election results, Franklin cried, "Oh, thank God for the good people!"

Paradise was also predominantly white, and in 1949, several years after the repeal of the Chinese Exclusion Act, there was opposition from the town's Chamber of Commerce to allowing a new Chinese restaurant to open, according to its proprietor. Yet The Pagoda, serving massive portions of chop suey and sweet-and-sour pork, eventually became a fixture and a venue for local community group meetings. The proprietor's daughter, Linda Lau, was one of two ethnically Chinese people at the local high school, along with her brother. To her surprise as a self-described nerd, she was elected the 1964 homecoming queen, beating out two cheerleaders. She wore a dress she had made herself in the school colors, white and green. "I was happy and a little shocked," she says. "I'm still baffled."

For all its faults, Paradise was a place where you didn't have to lock your door. It was a place where a man who described himself as the Paradise town drunk could sober up and be elected as mayor. And where the local paper's liberal columnist, who organized protests opposing Donald Trump's border policies, could become best of friends with his Trump-adoring neighbor.

The town hall, police department, and many of the businesses—Bobbi's Boutique, Dollar General, Needful Things & Antiques—straggled along, or just off, Skyway. Reflecting the relaxed attitude to planning that predominated before the town incorporated in

1979, the homes were scattered everywhere else: on roads that led off private driveways, in trailer parks, packed together on small lots, all obscured by a lush veil of vegetation. There was no sewer system, and everything relied on septic tanks.

In eastern Paradise was the town's largest employer, the Adventist Health Feather River hospital, a hundred-bed facility founded in 1950 by a group of doctors belonging to the Seventh Day Adventist Church. They took over 106 acres on the edge of a canyon for a sanitarium and an experimental farm. The land, with its fertile red lava soil, was ideal for growing "vegetables and fruits of especially high mineral and vitamin content as part of the hospital's therapy," the *San Francisco Examiner* reported in 1951. The sanitarium grew rapidly over the years, and a new hospital building was unveiled in 1968. By 1980, it had an emergency room and geriatric nursing program. In 2018, 1,500 people worked at the hospital and its affiliated clinics around the Ridge. It was an essential part of a community in which a quarter of the population was age sixty-five and older, and about 25 percent were disabled, more than double the rate in California at large.

Paradise residents were mostly middle- and lower-income, and it was a sanctuary for people priced out of the Bay Area or Southern California. In San Francisco a two-bedroom home went for upwards of $1.3 million. More than three hours away in Paradise, it was around $200,000. The Paradise Ridge was somewhere you could still buy your blessed, leaf-fringed patch of earth, your own little corner of the Golden State, and do whatever you wanted with it.

Above all, Paradise was the outdoors—the trees and the sun and the sky. Deer sunbathed in yards and bears tore through the garbage can if the lid wasn't tied. In the never-ending summer, families spent days camping by the water, fishing for crawdads, the beaches shimmering with fool's gold. They could day-trip up to the High Lakes or down to Lake Oroville.

A longtime resident, Don Criswell, who headed the local-history museum board, remembers "hot, black coffee beside a fire on a still-dark morning," the musty, evocative smell of ponderosa needles covering the ground, and the sounds of the Mamas and Papas echoing from car radios and handheld transistors through the forest. And there was the electric night he and his friends rode their horses to the Nelson Bar swimming hole and paddled atop the animals from one side to the other in the dark. Afterward they lay down to sleep, inhaling "the rich, lathery smell of wet horses, their blustery breathing and the smoke curling over us."

"Within a few minutes of leaving Skyway, all sights and sounds of civilization disappear." Criswell once wrote. "The soft whispering of wind high up in the treetops is there if a breeze happens to drift overhead. Anyone that's spent time in Paradise knows that sound, one of the most pleasant on earth, I think."

California's precontact Native Americans may have deployed fire as an ally, but it long ago came to be seen as a bogeyman in the US. Following devastating wildfires in 1910 in which seventy-eight firefighters were killed, the US Forest Service adopted an uncompromising approach. According to the so-called 10 a.m. policy of 1935, any wildfire had to be extinguished by that time the next morning. World War II, and the nightmarish visions of fire-bombed Dresden, Hamburg, and Tokyo, galvanized this attitude. Lives and homes were saved, but by the 1970s the 10 a.m. policy and its successor, the 10-acre policy, were revoked because the adverse effects of smothering every possible wildfire had become clear.

Fire suppression resulted in forests that would have been unrecognizable to Americans in the nineteenth century, and in wildernesses that were, ironically, much more flammable than before. Absent any flames, young trees and brush flourished unimpeded. In the late nineteenth century, photos of Yosemite Valley, for instance, showed large meadows. By the late twentieth century, these were filling in. The issue was not restricted to California—in some Arizona forests where previously there had been twenty trees per acre, there were now eight hundred. Timber companies and conservative politicians have argued that the answer is increased logging, but scientists have shown this to be the wrong approach, not least because it removes the largest, most fire-immune trees. Forestry officials have had some success adding fire back into the landscape with so-called prescribed burns, but once a given fire regime has become deregulated it is hard to reestablish. With people now living where fire previously flowed freely, fire defense costs are soaring, sucking up the majority of the US Forest Service budget.

The climate crisis threatens to lay waste to these efforts. Heat in California has crept up, particularly since the 1980s, and temperature minimums are now 2.3 degrees Fahrenheit higher than they were a century ago. The state's four warmest years on record were 2014 to 2017, and during the same period it experienced unprecedented drought. These conditions were connected with the deaths of approximately 129 million trees. In the Sierra, the snowpack, which keeps California more green and moist than it otherwise would be as the snow melts over the summer, is starting its annual ebb earlier than it has historically, and there is less of it. The rainy season is contracting, and the late arrival of precipitation in the fall months coincides with the time when vegetation is at its very driest, and when California is prone to high winds that can exacerbate a fire's spread.

Although there continue to be fewer fires overall owing to enormous investments in firefighting, the frequency of the most severe fires is increasing, because the new climatic conditions and the tangled state of forests and grasslands make those that survive initial attack harder to repress. Nine of the ten years of the most extensive fire activity in the United States have been since the year 2000, and by 2050 it is expected that as much as three times the acreage of western forests will burn as a result of global warming. In summer 2018, after ferocious blazes prompted California governor Jerry Brown to declare that the state had entered a "new normal," three fire researchers responded that this understated the crisis because "it would be a mistake to assume that the region has reached any semblance of a stable plateau." California, they said, had entered a new epoch of huge and fast-moving blazes—the "era of megafires."

Even though California's forests are adapted to fire, some researchers believe that the novel circumstances are becoming untenable. If a severe blaze wipes out a forest, there is a greater chance that the regrowth will shift from trees to smaller plants better adapted to harsher conditions. Certain tree species may manage to migrate upslope or northward, to their preferred conditions, perhaps with human help, even as chaparral and grassland plants such as manzanita, chamise, and deer brush gain new footholds. Ecologist Greg Asner, who has spent months performing aerial surveys of California's forests with a spectrometer and other instruments, calls this a "shrubification" of the state. "I picture a big shift, a big rearrangement of California, and I think it's under way right now," he said.

For Iris Natividad, Paradise represented safety, at least in the beginning. She and her partner of twenty-eight years, Andrew Downer, moved to town in 2015. It was their second time living in the area—they had left in 2004 for a lovely home in an out-of-the-way hamlet called Bangor, but the lack of cell service and the encroaching fires made them nervous.

Natividad and Downer were physical opposites: she a sparky five-foot-one DMV consultant who introduced herself as "Flower Christmas," he an irascible six-foot-six mechanic with a Father Christmas beard who told off-color jokes that cracked Natividad up. In Paradise they were able to indulge their shared hobby of buying and selling antiques.

"Paradise," Natividad said, "was an antiques destination." It was not that Paradise was on the same circuit as global contenders like London's Portobello Road. It was this: "Everybody was a senior. It was a gold mine." The preponderance of elderly people with a lifetime's worth of belongings meant a continually replenishing stock for the numerous antique, vintage, consignment, thrift, and collectible stores. You could spend a whole weekend trawling through them. Downer relished going to a store with Natividad and competing to see who could find the biggest steal—the $500 pitcher priced at $1, say. Their thing was glass, and their first love was uranium glass, which contains tiny amounts of the element and glows in the dark. It is not considered dangerous. Their home, a converted chiropractor's office downtown, was filled with their finds: plates and bowls of green Depression glass, a treasured large Heisey decanter in cobalt blue. Downer had 80,000 marbles in tubs, and there was a gumball machine filled with them on display outside the front door. Friends knew them as "fairies," the kind of people who would give you something of theirs if you expressed admiration for it.

Their home may have been a collection of everything they loved, but for Downer it was also a prison. He had had one foot amputated as a consequence of diabetes, the other barely worked, and he drank to comfort himself. Natividad worked away from home four days a week, from her office in the town of Santa Rosa, and while she was gone Downer had extreme difficulty even making it to the bathroom. He was terrified when she left, his counselor told Natividad, and she was frightened for him, too.

At a wedding party not long after Natividad moved back to Paradise, she met a firefighter and asked him about his job. How did he deal with the heat? Why did firefighters sometimes set small fires in order to extinguish larger fires? And also, were the fires getting worse? He knew they were, he told her. California was hotter and drier—it was climate change. "He actually opened my eyes to it," Natividad said.

This was on her mind when, the next year, the fires seemed to be chasing her. There was the LaPorte Fire, which destroyed seventy-four structures in Bangor. The Honey Fire burned 150 acres southwest of Paradise. And there was the Tubbs Fire, which killed twenty-two people in and around Santa Rosa. "In every one of those fires I know many people, and many friends, who lost homes," she said.

They agreed that Downer needed a vehicle in case he had to evacuate, and Natividad purchased an Infiniti SUV for him, a large car they thought he'd be able to maneuver himself into.

"It's not a matter of if," she thought. "It's a matter of when."

2

Off the Grid

At 10:20 p.m. on November 7, 2018, a Paradise public works official observed that all of the good work done by the town's street-cleaning program had gone to waste. Winds that had fomented in the oven-baked Nevada deserts were blowing over the top of the Sierra Nevada and coursing down past Paradise, scattering flammable leaves across town like confetti. The official made sure that emergency communications systems were connected to their backup batteries before calling it a night at 1:00 a.m.

Over the previous week, tens of thousands of Californians from the coast to Wine Country and the Nevada state line had received calls, emails, and texts from their utility, Pacific Gas and Electric, or PG&E, all with the same message: "Extreme weather conditions with high fire danger are forecasted in the North Bay, North Valley and Sierra Foothills, starting overnight Wednesday and lasting through Thursday. These conditions may cause power outages. To protect public safety, PG&E may also temporarily turn off power in your neighborhood or community. If there is an outage, we will work to restore service as soon as it is safe to do so. Please have your emergency plan ready. If you see a downed power line, assume it is energized and extremely dangerous. Do not touch or try to move it—and keep children and animals away."

Buried in the message was an unusual new policy born of bitter experience, which the company called "public safety power shut-offs." When the wind grew too strong and the humidity was too low—ideal conditions for the spread of a wildfire—PG&E would de-energize its lines until the danger had passed.

Residents bought dry ice and bottled water. At the police department, the new chief purchased a generator, and Rob Nichols, a fifty-two-year-old veteran officer, was girding himself for the week to come. When winds roared across the Ridge, as they often did, they rocked trees like piñatas and sent branches plummeting to the ground. During a storm years earlier, Nichols and his colleagues jokingly called Paradise "Windopolis." But such weather meant bad things for the town. Just a few years before a tree branch had fallen onto a car, killing a woman as she drove to church. Paradise was on the whole a great place to be a cop, Nichols thought—in nearly nineteen years, he had never once fired a gun on the job—but the Paradise Police Department was understaffed. It had fewer police officers than it did when it was first established in 1980, and there were 5,000 fewer residents, according to the police chief. A lengthy power outage in a town of 27,000 could mean people careening through intersections, and 911 calls when medical devices blinked off.

Paradise emergency coordinator Jim Broshears and his wife Cyd, a retired nurse, had taken extra care that week to rake the maple and pine leaves around their home, shed, and garage. They lived on a lot of almost an acre on Pentz Road, on the east side of town. The conditions warned of by PG&E were the kind Broshears dreaded. "That kind of wind tends to have a lot of potential," he said.

Broshears, a trim sixty-five-year-old with close-cropped gray hair, had hiked and explored these mountains his entire life, and knew how to read the sky and the landscape. Born in the city of Redding, 85 miles to the north, he had started working for Cal Fire two weeks out of high school. He found a job as an entry-level firefighter in Paradise in 1974 and had worked his way to the top: fire engineer, engine captain, assistant chief, division chief, and, finally, chief, from 1996 until 2006. The magnificent setting of Paradise suited him, and he hiked and fished from Butte Creek Canyon to Head Dam, the Little Pearl water hole and beyond. As fire chief, he also became the Paradise coordinator of emergency planning, a role he still held after his retirement.

Broshears had been instrumental in developing the town's formal emergency plan, and continued to update it through the years. Paradise was one of surprisingly few towns in California's fire-prone regions to even have a decent evacuation plan in the first place. The latest version, in 2017, split the city into fourteen evacuation zones. Evacuations would proceed based on which areas were affected, and during a disaster, cars would be permitted to use both lanes on certain roads to flee. Paradise sent out "know your zone" mailers to residents.

It also established a reverse-911 alert system, which would call residents with an automated message in the event of a disaster. But if they did not opt in to the program, they would not receive the warning. That was "the biggest drawback," said Broshears, who was involved with its rollout. "You are constantly trying to educate people to sign up."

Broshears spearheaded a number of projects intended to reduce the risk of fire. As fire chief, for instance, he pushed through a large brush-clearance project along the eastern boundary of Paradise, next to the hospital. He organized a drill in 2016 in which police and city officials staged a practice evacuation, shutting down stoplights at an intersection on Skyway and using volunteers to move traffic. A center turn lane was converted into an additional lane to help get cars out of town. And he had ambitions to use incentives and regulations to reduce the number of trees per acre in town by about two-thirds.

"We knew we'd have a fire," Broshears said. "We knew we couldn't escape it forever." Since 1999, at least thirteen large wildfires had burned in the region. Paradise had somehow always gotten off lightly. The Humboldt Fire, which began in June 2008 when an arsonist set a fire under power lines just east of Chico, burned more than 23,000 acres in and around Paradise. Heavy smoke and flames closed three of the four main evacuation routes out of town, and the fourth, a two-lane road, was so congested that what should have been a fifteen-minute journey south lasted nearly three hours. Ten firefighters were injured and eighty-seven homes were destroyed, yet Paradise avoided the worst of it. Most of the damage was immediately south of town, and only six of the lost homes were within its boundaries.

An editorial in the *Paradise Post* emphasized how lucky the town had been. "If last week's fire did anything it brought home to every last Paradise and Magalia resident the vulnerability of our perch here on this lava ridge. At 36 square miles in size [twice the size of Paradise itself!], at times moving faster than anyone could run, this historic marauder of fire, wind, and smoke swallowed everything before it in a three-day rampage across the southern peripheries of town." It surmised: "The real damage to those of us who live in town was to our confidence and, certainly, our complacency."

The same day the piece was published, the BTU Lightning Complex Fire was sparked by lightning in the mountains north and east of Paradise. It burned nearly 60,000 acres and destroyed more than one hundred structures. Yet as had happened on previous occasions, the West Branch of the Feather River, in the steep

canyon that formed the town's eastern boundary, was a bulwark
to the encroaching flames. "How in the world does a fire burn to
the waterline for 5 miles and not cross?" Broshears said. "It just
blows my mind." According to a county grand jury that considered the incident, it was only "by some miracle" that the flames
halted at water's edge. If they had not, "property damage could
have been huge and thousands of lives could have been threatened
in Paradise and the Upper Ridge."

The grand jury recommended the improvement of evacuation
routes to the north and south, the removal of trees and brush,
and a moratorium on multihome development in fire-prone areas
until "fire safety, traffic, and emergency water supply issues" were
resolved. Though concerned about safety, some county officials
believed the moratorium was unreasonable and said that building-
code improvements made them unnecessary. "There are systems
in place to address these issues," the county board of supervisors
wrote in response.

Inhabitants of the Ridge later learned of the perils facing even
sizable California towns. In July 2018, ash from a fire 80 miles
away had fallen over Paradise. Named the Carr Fire, it had started
on a 106-degree afternoon in July, when a trailer driven by a couple outside Redding, a city of 92,000 people, had gotten a flat tire
and its rim scraped against asphalt. Sparks had ignited a fire along
the highway that jumped into a 42,000-acre unit of the National
Park Service, consuming miles of dry brush, growing quickly in
size and speed, and sweeping toward town.

The Carr Fire killed eight people, burned for thirty-nine days,
and destroyed almost 1,100 homes. But it captured the world's
attention for another reason: it birthed a remarkable fire tornado,
a product of the high temperatures and hot air from the fire creating a column of superheated air that began to rotate as it rose. The
twister was 1,000 feet wide and tore through the area, uprooting

trees, tearing roofs off houses, and throwing power lines and cars into the air. Whirling at 143 mph, it reached temperatures of 2,700 degrees and heights of 17,000 feet. A firefighter died when the tornado snatched his 5,000-pound truck and flipped it down a road.

Redding's Republican congressman, Doug LaMalfa, had previously expressed the view that "there's a lot of bad science behind what people are calling global warming." Remarkably, in the face of the tragedy in his own district, he still declined to make a link between wildfires and global warming. "I'm not going to quibble here today about whether it's man, or sunspot activity, or magma causing ice shelves to melt," he told a reporter, referencing debunked alternative theories for hotter temperatures.

The *Paradise Post* had published regular updates on the fire, and Ridge residents gathered donations of snacks, water, clothing, and baby diapers and formula to deliver to the besieged area. When a resident complained on a community Facebook page that there was too much coverage of the disaster on local news, others jumped in to say: *That could be us.* In the preceding two years alone there had been at least three fires around Paradise, one caused when tree branches fell onto a PG&E powerline.

"A fire can't do what the Carr Fire did," said Broshears. "It's just not possible." Like other firefighters, he was confronting a reality for which he had no reference point. In 1972, he'd fought what was considered a "major fire"—the Shasta Fire, which had prompted a couple thousand people to evacuate and had eaten up 1,700 acres. The Carr Fire, by comparison, claimed 230,000 acres. The footprint of the Shasta Fire "in the footprint of the Carr Fire is a dot," he said. "It's just a *dot.*"

If the worst-case fire were to hit Paradise, Broshears had an idea of what it would be like. He considered the town vulnerable to a wind-driven fire, the kind that invaded the San Francisco Bay Area in 1991. The Oakland firestorm began on a Saturday afternoon in October with a grass fire in the Berkeley hills. Fire-fighters extinguished the blaze—or thought they had—but by the next morning it reignited and, driven by winds in excess of 65 mph, it quickly grew. Overwhelming firefighters, it spread to nearby homes, and leapfrogging embers started new blazes in the distance. The Tunnel Fire, as it became known, jumped a four-lane and an eight-lane freeway, moving so quickly that many residents had little or no warning. It killed twenty-five people, most of whom got stuck on narrow, winding roads as they tried to flee. Almost 4,000 homes were destroyed.

"I certainly thought an Oakland Hills could happen to us," Broshears said. When people asked him why Paradise had so far avoided that fate, "I didn't even go down the list of all the great things we'd done. I said, 'Luck. We're lucky a fire hasn't burned into town.'"

Broshears was ready for a possible power shutoff—he had battery-powered lights and a gas stove. Outages weren't uncommon on the Ridge. During a period of heavy snowfall in the early 1990s, the electricity was off in town for two weeks. Other locals, of course, were concerned and annoyed. "PG&E should be working to prevent the need to turn off power, by fixing obvious issues before they become threats," one woman posted in an online community group on November 7. Another user wrote that night: "It's better than the town being ravaged by fire."

PG&E transmission tower 27/222 had stood in the deep, forested canyon for almost one hundred years. It was 7 miles east of Paradise as the crow flies, separated from the town by a hilly plateau on which few people lived, except for the 800-strong community of Concow, located midway in a circular valley. The tower suggested a human form, balanced on stout, steel legs that supported a trunk and outstretched arms festooned with cables.

Part of the 56-mile Caribou-Palermo transmission line installed between 1919 and 1921, the lattice tower had outlived the people who built it and the company that first owned it, and was still in operation almost twenty-five years past what PG&E called the "useful life" of such structures. It had planned to upgrade the line, but said that engineering challenges and problems obtaining permits for segments that run through federal forests had made it difficult to do. PG&E crews had not climbed the tower to inspect it since at least 2001.

Yet problems were apparent. A 2012 winter storm with winds of up to 55 mph knocked down five towers on the same line as 27/222. PG&E temporarily supported the line with wooden poles. In 2015, the California Independent System Operator, the nonprofit tasked with overseeing the state's electric system, identified "several reliability concerns" and a need to improve and upgrade the power system serving far Northern California to prevent outages and thermal overloads, which occur when there is too much power moving through a line.

Electrical utilities are hardly household names beyond the actual homes of their customers, but PG&E is different thanks to its role in the film *Erin Brockovich*, which focuses on the company's culpability for groundwater contamination in the Mojave Desert town of Hinkley, California. Wastewater from a PG&E facility had leaked into Hinkley's groundwater, infusing it with the carcinogen hexavalent chromium. Residents sued the utility and won a $333 million settlement. Today, PG&E is the largest electricity provider in California. It serves 16 million people in a 70,000-square-mile service area with power from gas, hydroelectric, nuclear, and solar plants, all threaded together by more than 125,000 miles of power lines. It is a monopoly, albeit a regulated one, overseen by independent state regulators. And it is owned by investors, to whom it is expected to deliver a profit.

Tower 27/222 was part of an electricity unit of which PG&E was particularly proud—a set of eight hydroelectric powerhouses that are located on the 70-mile North Fork Feather River, one of four tributary forks of the main Feather River. They are capable of generating 690 megawatts, enough to power hundreds of thousands of homes. In the 1940s the company showcased one of them, the Caribou Powerhouse, to distinguished guests. Oddly enough for a power plant, it had what the company described as a "comfortably appointed clubhouse" and, until 1950, a skilled Chinese chef. PG&E had a name for this little hydroelectric network, evoking the river's rapid descent through the hills, and suggesting its influence on the state at large: the Stairway of Power. Its story helps explain how California became so uniquely, and combustibly, electrical.

California has been dubbed a "hydraulic society" by environmental historian Donald Worster because, like ancient Mesopotamia and Egypt, the social order is "founded on the intensive management of water." Precipitation is unequally distributed in the state: most rain and snow falls in the north and east, while the coastal metropolises of the San Francisco Bay Area and Los Angeles are far drier and verge, in the latter case, on desert. To move water from where it was abundant to where it was craved, the government engineered a web of pipelines, canals, and dams of extraordinary complexity. Yet California might also be called an "electrical society," because to enable its early growth it tapped those same far-flung water supplies—not for drinking or irrigation but, in the absence of other options, for power.

In the late 1840s, at the dawn of the Gold Rush, "San Francisco walked in darkness," Charles M. Coleman wrote in a 1952 history of PG&E that it had commissioned. Light was provided by candles made of tallow or whale oil, for which there were four refineries in the city at the time. California had few deposits of coal, the primary fuel in the eastern states. In the goldfields, however, miners were developing expertise in working water. To obtain access to the gold in stream beds, they diverted the flow into artificial channels, and could subsequently use the water at high pressure as a powerful substitute for a pickax. By 1870, they and water companies had built an estimated 5,000 miles of ditches, and soon energy companies took an interest in them. "The air of the whole Pacific Coast has all at once become filled with talk about setting up water wheels in lonely mountain places and making them give light and cheaply turn other wheels in towns miles away," the *San Francisco Call* wrote in 1895. By 1900, there were twenty-five hydroelectric plants across the state, and because they were often situated so far from where the power was desired, California was a pioneer in the long-distance transmission of electricity. The countryside was strung with lines that continually broke records, such as the 4,427-foot line that crossed the Carquinez Strait in San Francisco Bay.

Electricity—invisible, clean, arriving magically from afar—was glorified. PG&E began as a Gold Rush–era gas company that diversified into hydropower, and in 1916, it paid for new street lighting in San Francisco. Seventy years earlier the city had been submerged in rank darkness, but now "a warm white light, the most brilliant that ever shone through a city thoroughfare after the sun had gone down, flooded the canyon of Market Street," wrote the *San Francisco Chronicle*. "Light flowed everywhere, touched and enveloped everything . . . such was the wizardry of this marvelous light."

So bountiful was this energy from the mountains that soon each Californian used 2.5 times more electricity than people back east, and hydroelectricity made up over 80 percent of California's power supply until after World War II. It thrummed through everything from the cement mixers that helped rebuild San Francisco after the 1906 earthquake to the irrigation pumps that were helping make the Central Valley into the agricultural envy of the nation and the studios of the nascent motion picture industry in Los Angeles. Westerners know how to "do it electrically," one journal bragged. Hydroelectricity even spurred the consumer culture for which California became renowned. In order to boost usage, the utility Southern California Edison sent out salesmen with household appliances. In 1911, for example, it sold 15,438 flatirons and 4,634 coffee percolators.

PG&E developed a massive hydroelectric network that spanned the Sierra Nevada, though drought in the mid-twentieth century, and the resulting weak river flow, prompted efforts to diversify and cheap oil and gas made steam-powered plants a more attractive proposition to utilities. By the 1950s, PG&E supplied power to forty-six of fifty-eight counties in California and Coleman distilled its essence, as he saw it, into an encomium to its workers: "If one seeks to define PG&E, he may find it in the company's 17,000 employees—its ditch tenders and lake tenders; its linemen, who risk their lives and endure hardships that service may be maintained without interruption; its operators who stand lonely watches in isolated powerhouses to hold constant the flow of power from the generators; its servicemen whose skill is on call for the aid of customers when gas or electric appliance or motor operation must be restored."

But as PG&E expanded through the latter half of the twentieth century, constructing a 612-mile pipeline to bring gas from Canada to Northern California and opening a nuclear plant that was later found to be sited almost directly above a newly discovered earthquake fault, a fresh set of risks emerged. In the early 2000s, PG&E teetered on the brink of collapse after California attempted to introduce competition into the energy sector by encouraging utilities to sell off their power plants and trade electricity on the open market. A shortage of electricity, as well as manipulation by companies like Enron, led PG&E to pay far more for power than it could afford, and it declared bankruptcy with $9 billion in debt.

After restructuring, PG&E rebounded. Its annual revenue grew from $11 billion in 2004 to $17 billion in 2017. And as ever more people moved into the wilderness, it followed them there with ever more power lines. These lines have proved a deadly weakness. Ordinarily around 1,000 PG&E linemen were tasked with maintaining them, climbing up towers as high as 300 feet or dangling over the massive structures from helicopters. By most accounts they are diligent and dedicated to safety. They work long hours, missing Christmases and birthdays, often in isolated corners of the state, hiking amid chaparral rattlesnakes or driving snowcats into frozen mountains.

Another 1,500 or so contractors had the grueling job of "vegetation management." PG&E is required to cut down any trees or bushes encroaching on its lines, and maintain clearance in case a fire is sparked by trees falling on power lines or vice versa. In the wake of a particularly deadly fire season in 2017, PG&E's regulator, the California Public Utilities Commission, or CPUC, implemented new vegetation and wildfire safety standards, requiring power companies to maintain clearance of 4 feet around power lines in high fire risk areas, and recommending a minimum of 12 feet in order to ensure compliance with the regulation year round as vegetation regrows.

The consequences of failure are devastating. Wildfire ignitions from power lines are the only kind that have increased in California since 1980. PG&E equipment was found to have caused at least 1,500 fires between 2014, when the state first began requiring the utility to report such blazes, and 2017. Most of these were small and caught quickly, but in recent decades there is a grim and surprisingly long litany of those that were not.

In 1994, for instance, a fire destroyed twelve homes and twenty-two structures, including an 1868 schoolhouse, in the nineteenth-century town of Rough and Ready in the Sierra Nevada. It began when a power line made contact with a tree limb that PG&E should have pruned. During random inspections, investigators found several hundred other safety violations, including nearly two hundred instances of vegetation contacting power lines, in the vicinity of the blaze. A jury convicted the company of 739 counts of criminal negligence and required it to pay $24 million in penalties. After the trial, a report by the state utilities commission revealed that PG&E had used $77.6 million from its tree-trimming budget for other purposes in the preceding years, and neglected to spend $495 million that had been designated for maintenance.

Cases multiplied. The 1999 Pendola Fire burned 11,725 acres partly in the Tahoe and Plumas national forests after a rotten pine, which the federal government found PG&E should have removed, fell on a power line. The company settled with the US Forest Service for $14.75 million, and with the public utilities commission for $22.7 million, after regulators determined that PG&E was failing to spend what had been allotted for vegetation management. The Freds and Sims fires in 2004, which each burned more than 4,000 acres of federal forest land, were also found to be caused by trees falling on company power lines. PG&E again paid out millions in settlements.

The fires have turned deadly. On September 9, 2015, a blaze broke out about two hours northwest of Yosemite National Park when a tree came into contact with a powerline. It sparked a massive fire that consumed 14,500 acres over the course of a single day, killed two people, and destroyed 550 homes, the majority of which were in Mountain Ranch, a tiny Gold Rush town of 1,800. News reports described Mountain Ranch as being nearly "burned off the map." It emerged that the 44-foot gray pine that started the fire had been pinpointed as a "hazard tree" that could fall into PG&E's lines. State regulators fined PG&E $8.3 million for failing to maintain its lines and the clearance between its lines and vegetation. Cal Fire, the organization that fights wildland fires in the state, sent PG&E a bill for $90 million to cover fire-fighting costs.

How could one company be linked to so much catastrophe? In 2010, after yet another disaster, a gas pipeline explosion in the town of San Bruno that killed eight, PG&E's inner workings were heavily scrutinized. A 2011 investigation commissioned by the state found a "dysfunctional culture" that was reactionary rather than founded on long-term safety planning, that the utility's "rhetoric does not match its practices," and that its safety goals were disconnected from what was happening on the ground. Another audit for the California Public Utilities Commission found that PG&E deferred maintenance projects to increase its profits.

Critics of the utility contend that while each accident is complex and distinct, one issue undergirds them all. "They're a shareholder-owned, profit-making company," said Ken Pimlott, who was the head of Cal Fire from 2010 until 2018. "How do you take out any implication that what they're doing out there is driven by a profit motive versus purely trying to address public safety? How do we know they're investing everything they can in mitigating thousands and thousands of miles of power line structures?"

PG&E has maintained that it can only do so much in the face of the climate crisis. A Northern California lineman who worked for the company for twelve years said PG&E was doing its best to accomplish a Sisyphean task. "It's an aging infrastructure. There's millions of miles of power lines," he said. And while PG&E workers are grieved by accidents involving the company's equipment, "they understand that utilities don't operate in a laboratory," said Tom Dalzell, business manager at the International Brotherhood of Electrical Workers Local 1245, the union to which the largest number of PG&E staff belong. "They think that in general, the mistakes that are cited are part of having a utility."

Outside experts acknowledge the difficulties facing the company. "Even if PG&E is much more aggressive about trying to harden its infrastructure and makes its system more safe, it's going to take a tremendous amount of time, probably decades, to cover all their equipment in meaningful ways," said Steven Weissman, who oversaw PG&E as a former judge with the California utility regulator. And under a California judicial doctrine, utilities can be deemed legally responsible for wildfire damage caused by their equipment regardless of whether they acted negligently, and even if they followed state safety rules. This makes them vulnerable to expensive lawsuits, the cost of which may end up being covered by shareholders or ratepayers, despite how galling this may seem to the latter.

PG&E has taken steps to mitigate fire risks, establishing a wildfire safety operation center in San Francisco, which is staffed with meteorologists and other experts and open 24/7 during the fire season. It had added about a hundred of its own weather stations by September 2018. And between 2014 and 2018, the company provided more than $13 million to local fire safety councils.

PG&E took a new approach after the cataclysm of October 2017, when over a dozen fires ignited in rapid succession on a Sunday night and roared through Northern California. The deadliest, the Tubbs Fire, started when dry winds at racehorse speed drove flames from the Sonoma County foothills to Santa Rosa, a suburban city of 175,000. Residents in Coffey Park, a 1980s development of single-family homes, awoke to smoke-filled rooms and walls of flames surging toward their homes. By the end, forty-four people, including twenty-two in the Tubbs Fire, and as young as fourteen, were dead. Early reports indicated PG&E was responsible, and the company expected to pay $2.5 billion in liabilities. In the wake of the fires, it announced its power shutoff policy. Though a state investigation ultimately found that PG&E was not responsible for at least the Tubbs fire, the policy remained in place.

It initiated its first prophylactic shutoff in October 2018, affecting 60,000 people north and northeast of San Francisco for about a day and a half. Although it was meant to protect them, customers weren't happy about it. In editorials and public meetings, they wondered how the elderly, disabled, and those with medical devices such as ventilators would manage without power. To some, it seemed too easy a way to shrug off responsibility for maintenance and safety, even though customers were presumably already paying PG&E, in the form of their electricity bills, to get its ship in order. "It put the onus on all of us, the public at large, saddling us with the loss of hundreds of dollars from the loss of food in our refrigerators and freezers," one wrote to the *Santa Rosa Press Democrat*.

Once risky conditions have passed, restoring power is not as simple as flicking a switch. Utility workers must inspect lines in the shutoff area to ensure the wind hasn't brought down branches or vegetation that would spark a fire when electricity is again sent zipping through the system.

In the first week of November 2018, there were still pumpkins on porches in Paradise. The town hall got an upgrade after a councilwoman donated her monthly stipend to pay for paint and a new sign. The midterm elections were the talk of the town—unlike in much of the country, there was no Blue Wave—and there were local races, too. The owner of a local embroidery shop promised to tackle "transient" issues, drug use and theft on the ridge and duly made it onto the council. A sign at the Ace hardware store advertised a Christmas event with Santa for November 8.

It was achingly dry. The area hadn't seen a decent rain since spring. Chainsaws had been buzzing out in the woods as PG&E carried out its new policy of ensuring 12 feet of clearance between vegetation and powerlines. That week, the company warned Paradise that the outage could last as long as seven days, but that it would have 135 workers and as many as eight helicopters inspecting the lines to get power restored as quickly as possible.

"This will be an inconvenience for everyone," the town warned its residents on social media. "We need to be prepared."

The possibility of disaster lingered on some people's minds. At a candidate forum for the town council election a couple of months earlier, moderators had asked five council candidates whether the town was prepared for an emergency such as a wildfire. The mayor, Jody Jones, was confident. A retired planner with the California Department of Transportation, she was elected to the council in 2014, and led modestly and methodically. The town had a system, had staged practice evacuations, and residents knew their zones, she said. It was a good plan—the town just needed to stick to it.

Julian Martinez, a council candidate whose signs described him as the "working class choice," had offered a different view. He said Paradise was equipped to handle the fires of ten years ago, not the tornado-spawning, suburb-razing fires of today. "We are prepared for minor disasters regarding wildfires. But in a worst case scenario I don't think we are. The amount of overgrowth that's inside of this town is dangerous. . . . I've been close to fires before and I know how they move through the canyons with the wind. They can catch you off guard really fast. We've got a lot of elderly people in this town, we've got a lot of people who don't have cars, or even worse, cars that aren't reliable."

In the early hours of November 8, the weather on the Paradise Ridge became treacherous. Normally the wind in California blows from west to east—from the Pacific Ocean inland. But, particularly in the fall, a different pattern emerges. Dry, northerly air streams in from the Great Basin. Such warm downslope winds, known in different parts of California as Santa Ana or Diablo winds, are feared by firefighters because they disperse wildfire embers as easily as a child flinging dandelion seeds into a breeze.

Gusts upwards of 50 mph swept down over the foothills. A sixty-six-year-old retiree named Susan Van Horn had trouble sleeping that night. Lying in bed in her double-wide mobile home, she could hear gusts tearing through the oak trees on her half-acre property, and falling branches and acorns clattering onto her roof. "I thought, 'If we're going to have a fire, this is it,'" she said.

Rob Nichols, the police officer, was also awake. Wind was shaking the last remaining trees on his property. He had cut down six oaks the summer before, in part because he worried limbs might fall onto the home he shared with his wife and two young children, and in part because they were messy, constantly dropping leaves. It was around 4:00 a.m. and the glow from a nightlight told Nichols the electricity was still on.

A PG&E team was weighing its options and monitoring weather conditions to decide whether to implement a power shutoff. The company ultimately elected not to. Ironically, even if it had proceeded, tower 27/222 would have remained energized. PG&E's shutoff program included only smaller, lower-voltage distribution lines, which transport energy locally, and not the massive transmission lines responsible for supplying power to entire regions and cities, such as the Caribou–Palermo system.

On its hill, tower 27/222 was emitting its usual low thrum in the violent darkness. Sometime after 6:00 a.m., a hook high up on the arm of the tower broke, releasing an electrical wire, and sparks scattered into the brush below.

3

Firebrands

Four miles east of Paradise as the crow flies, fire chief Matt McKenzie awoke in confusion at 5:00 a.m. on Thursday, November 8. It sounded like rain hitting the metal roof of the fire station, but there was none in the forecast. It took him a moment to realize that it was pine needles being flung by a strange wind—not the usual intermittent gusts he was accustomed to hearing swoosh down the canyon, but a sustained, jet-engine roar. He got up and went to the station's small kitchen, where he put the coffee on and started dicing potatoes and onions for his crew's breakfast.

A twenty-year veteran of Cal Fire, McKenzie had never rested easy at Jarbo Gap. Its location alone was enough to quicken the heart. The station was atop a high ridge overlooking the canyon in which the North Fork Feather River flowed southward, at about the point the canyon and the river make an abrupt, 90-degree turn to the east. Winds streaming down the canyon would run straight into this turn and spill out, moaning over the top of the station, especially at night, when canyon winds commonly pick up. Fires in Butte County often start in this steep and inaccessible declivity, served by a single main road that meanders alongside the water.

At 6:15 a.m., PG&E experienced a power outage on the Caribou-Palermo line. Only one customer was affected—a hydroelectric station located in the canyon that was operated by a small city in the San Francisco Bay Area. At about 6:28 a.m., a PG&E hydro generation supervisor driving through the canyon six miles north of Jarbo Gap saw a fire that he estimated to measure 100 square feet in a clearing below the transmission line, and radioed a nearby company facility, which conveyed the alert to emergency services. Before long, McKenzie's phone lit up with a text informing him of an ignition.

McKenzie opened the back door to the kitchen and the wind ripped it right out of his hands. Although he expected to catch the smell of the fire, all he got was the dry pines. He and five other firefighters headed into the canyon in the dark in two trucks. It was a place McKenzie had known his entire life—as a kid growing up in the nearby town of Oroville, he would come up here with his dad to hunt, fish, or log a Christmas tree in the high, green solitude. It took them about fifteen minutes to reach the fire, going as fast as they could while keeping the engines on the road and avoiding nosediving into the canyon. As soon as they got there, McKenzie was dismayed.

The flames—sheltered and low to the ground, a dusky orange—were on the wrong side of the river. McKenzie had been hoping that they would be on the canyon's east side, where they would be accessible from the two-lane highway he was on. Instead they were west of the river, off a dirt track called Camp Creek Road. It was an impossible location. Camp Creek Road was hewn roughly into the side of the valley and in some places had been washed out by floods. It was so narrow that a fire truck could pull in its wing mirrors and still probably scrape the rock outcroppings. It could take another forty minutes just to carefully creep down to the blaze, and once they reached it there would be nowhere to turn around, meaning the truck would have to reverse out. That would be agonizingly difficult under any circumstances, but especially in the face of flames being wind-whipped toward them. It was a situation in which McKenzie would want to call in aircraft to drop water or retardant, but he knew that wasn't an option: it was too dark and the wind too savage. "It was so close to a highway yet so far away," he said. "It was taunting—you see it, look at it, and it's right there, and you can't do a frigging thing about it."

Camp Creek Road may have defied fire crews, but it did lend the blaze its incongruously playful name: the Camp Fire. Analyzing from the highway, McKenzie thought the wind was behaving unusually. Typically it blows south, toward his station. This time it was blowing west, as if trying to push the fire up the wall of the river valley. This was why he had been unable to catch its smell earlier. It looked to him like a fire with enormous potential, the kind of fire he might want to throw hundreds of fire engines at. But there were only a fraction of that number in Butte County. He radioed the incident command team in the town of Oroville, the county seat, and requested fifteen additional engines and four bulldozers. His hope was that they would catch the fire once it topped the slope it seemed to want to crest.

That morning, many of the senior Cal Fire commanders in Northern California were at a leadership meeting that took place every month or two, and were staying at a hotel by the 101 Freeway in Marin County, about 130 miles southwest of the ignition point. Cal Fire is responsible for fire protection across more than 30 million acres of California wildland, excluding cities and areas that might have their own fire departments. The century-old organization acquired its first fire truck in 1929, created a "conservation camp" to train inmates to fight fires in 1946, and began using converted World War II bombers as air tankers in the 1950s. It is one of the largest firefighting agencies in the US, and has the biggest firefighting air force in the world. Boasting more than 800 fire stations and more than 5,000 full-time workers, it responds to more than 5,000 fires a year, and maintains a fleet of tankers, helicopters, and bulldozers.

Ken Pimlott, the head of Cal Fire, was awakened at 6:30 a.m. by a text from one of his senior commanders. *This is a bad area*, it said. *I've got concerns.* Pimlott knew the region. He had climbed the ranks since becoming a reserve firefighter in Contra Costa County, across the bay from San Francisco, right out of high school. Cal Fire later dispatched him across the state—he spent ten years in Southern California, where his roles included fire chief in the small town of Moreno Valley; he became a division chief in the northerly El Dorado County, and held influential positions at headquarters in Sacramento. He took the top job just after Governor Arnold Schwarzenegger left office and was succeeded by Jerry Brown.

Rather than having its own fire department, the town of Paradise had contracted firefighting services out to Cal Fire, like many smaller communities. On one visit Pimlott had noticed with concern the town's heavy tree canopy, the narrow roads, the dense settlement.

This meeting was supposed to be Pimlott's last before he retired from the agency, and by this late in the year the fire season would, in decades past, have been on the verge of concluding in Northern California. It traditionally ended with the onset of the winter rains. But over the past decade, with blazes sparking in December and January, top firefighters began to wonder whether there was even such a thing as a "fire season" anymore. And years of fires of ever-mounting ferocity had shown Pimlott that all bets were off. Who imagined you could top the 2013 fire that tore into

Yosemite and burned through hundreds of miles of mountains? But then came the 2017 devastation in Santa Rosa, the 2018 fire tornado, and, also that summer, the largest fire in recorded California history, the Mendocino Complex Fire, which burned 717 square miles in the state's northern reaches, an area more than half the size of Rhode Island.

Meanwhile, costs for firefighting in California rose from $90 million in 2010 to $947 million in fiscal year 2018, which ended in June 2018. That summer, the state fire agency had to request additional funding after it spent $432 million in just two months while fighting the Mendocino Complex Fire. "It's almost like you get numb, because every year we were topping the previous year," Pimlott said.

Although Pimlott had oversight of the agency, it wasn't up to him to direct the details of the Camp Fire response. Cal Fire operates on the principle that whichever officer is first on the scene, be it a first-year engineer or bulldozer operator, assumes command of the incident, orders resources such as engines, bulldozers, and aircraft, and sets mission objectives. A senior officer subsequently takes over and directs additional firefighters reaching the blaze. In the most chaotic fires, when the disaster is unfolding too quickly to get a grip on it, the response may operate under a principle called "leader's intent." In these rare, sobering scenarios, firefighters coordinate with each other via radio but have significant latitude to decide how they will go about saving lives and buildings. It had happened the year before, during the Wine Country blazes. "It's organized chaos," said Pimlott. "It's very much a war zone."

Matt McKenzie, the fire chief at Jarbo Gap, was the first incident commander, though eventually a local battalion chief took over and bolted up to McKenzie's location to see for himself. The mission objective was, at first, to restrict the fire to the area east of Paradise.

News of the conflagration spread quickly. Down in the valley, about 50 miles away, Chris Haile was awake early and logged on to Facebook. A former Cal Fire division chief in Butte County, Haile liked to keep an eye on conditions. *"Batten down the hatches!"* he wrote to a friend in the hills as the wind raged at 6:00 a.m.

Pete Moak, a fourth-generation local logger and rancher who lived in Concow, replied that the weather did not bode well: "We are running on borrowed time." Moak and his wife, Peggy, had a comfortably appointed, rustic home equipped with a generator, water tank, and two full pantries. Moak's knowledge of the terrain and its propensity to burn was second to none, and he didn't think Concow was a suitable home for people who lacked the means to look after themselves. His grass was always green, because a well-watered lawn is a firebreak.

Just past 7:00 a.m., Haile sent a message to Howard Goodman, the only inhabitant of a pinprick community north of Concow called Rag Dump: "Fire in the Camp Creek area, Howard. It has a lot of potential."

"This is not good," Haile posted. "Bad place for a fire."

When Haile uploaded a picture of the plume that he had just taken, Goodman responded:

"That's steep terrain filled with bush, and it's like a wind tunnel."

"I fought some fire in there in my time, don't have any fond memories," wrote Haile.

Another acquaintance chimed in. "The smoke is so thick in Paradise, we thought at first the fire was here."

Goodman announced that he was taking a drive up Flea Mountain, a 5,000-foot peak where he could get a good vantage point. Shortly after, he posted a warning:

"Concow, it's coming for you, evaluate."

Fire engineer Elliot Hopkins arrived at his home station of Jarbo Gap about forty-five minutes after McKenzie had left, for his first day back at work after a vacation. There was just one fire engine there, and a buddy who told him there was a ripping fire down in the canyon before racing off on the truck.

Hopkins, a strapping twenty-seven-year-old, couldn't see any sign of flames and wondered what to do with himself. Five minutes later, a senior Cal Fire training officer named Jeff Edson bustled into the station in his white department pickup to fuel up and told Hopkins to jump in. They were headed to Concow.

Named for the Konkow people who had been hounded to the brink of disappearance, the community centers on Concow Reservoir, which presents as an alpine lagoon fringed with tule reeds and piney mountains. The valley is a dimple in the gnarly uplands that separate Paradise and Jarbo Gap, and despite its relative proximity of only a couple miles to both, the circuitous roads mean that it takes forty-five minutes to drive to the former and twenty-five minutes to the latter.

People moved here to work the land and for the sun-speckled, dirt-road lifestyle. In the early 1970s, Concow's population of loggers, homesteaders, and retirees was supplemented by hippie types looking for a more sustainable kind of existence, inspired by the *Whole Earth Catalog*, Wendell Berry, and the back to the land movement. A commune called the Mountain Family was established, and in 1975 the long-running craft and music festival Wild Mountain Faire began. It managed to bring together both the tie-dyed crowd and Concow old-timers. The valley also had a "guerrilla marijuana culture," said Mike Ashlock, who moved from the Bay Area in 1983 or so to join the Mountain Family. To provide income in a place with few jobs, and for their own use, residents grew weed in discreet corners—screened by trees on their own property or in the lee of bushes on Forest Service land—so it wouldn't be spotted from law-enforcement helicopters. After California legalized marijuana, one man openly cultivated pot plants of such dimensions that he had to remove the roof of his greenhouse to accommodate them.

Taking the tortuous Concow Road that led into the hidden valley, Hopkins and Edson saw an ominous sign. The pillar of smoke that pours into the sky above a fire is known as a convection column. Composed of carbon dioxide, water, particles of burnt matter, and thousands of other components, a convection column can reach as high as 45,000 feet under certain weather conditions and create its own clouds, known as pyrocumulonimbus clouds, and thunderstorms. But Hopkins and Edson noticed that the winds hitting the Camp Fire were so strong that they were pushing the column nearly horizontal to the ground. Anyone caught inside it would be subject to its nastier effects, from particles clogging their airways to formaldehyde and acrolein irritating their eyes and lungs. Concow was about to be choked in a gritty, black shroud.

Edson told Hopkins to get on the loudspeaker: *Imminent fire danger*, he proclaimed. *Evacuate immediately*. Even so, they couldn't quite believe it when they suddenly came across flames—a 10-by-10-foot blaze burning near a home, and a man outside trying to tamp it down. In severe fires, it is common to encounter "spotting," when burning leaves, pine needles, or bark are blown ahead of the main body of the fire, like so many million flying matches. These incendiary chunks of flaming debris are called firebrands, and the new fires they spark are known as spot fires, which can create a vast lattice of firelets that threaten to merge and lay waste to a firefighter's best-laid plans. As far as Hopkins knew, spot fires tended to occur where the main fire front was at least visible—but the Camp Fire was still miles away. Edson radioed for twenty-five engines to hustle to Concow to help evacuate residents and protect buildings. And the men told the homeowner they couldn't do much for him—they didn't have a hose or tools on their truck.

Concow Road reaches the lake at its southern tip, follows the eastern shore of the roughly oval-shaped body of water, and at its northern tip it intersects with Hoffman Road, which leads west, leaving the lake behind and heading toward the hills between Concow and Paradise, and a landmark named Sawmill Peak. Edson and Hopkins saw five spot fires burning on the slopes of the mountain, all about 5 acres in size. Receiving a report of a trapped person, they turned around and headed back to the intersection. It didn't seem that more than a few minutes had passed when the vegetation on either side of the road was catching fire. In the middle of the street they found a woman in her pajamas yelling that she didn't have a car and couldn't get out. Hopkins and Edson pulled her into their pickup; she sat behind them.

On lots hidden by screens of trees all around them, frantic decisions were being made. Long-timers fired up their tractors or chainsaws and began clearing flammable brush from around their properties, to create a swathe of defensible space. Others jumped into their cars and poured out onto bad roads on which visibility was rapidly diminishing.

Like many of her neighbors in Concow, Joanna Curtin was off the grid. She was a woman in her sixties who lived without electricity in a mobile home on 160 acres. She was born in Malaysia to English parents, attended school in the UK, and as a twenty-one-year-old met an American stationed at an air force base near the town of Ipswich. They moved to the US together, but he died after falling from a military truck, and she'd come alone to Concow, a place she found bewitching. She raised two daughters and now rescued horses intended for slaughter, and friends often gathered at her place for long evenings of drinking wine and playing music in the lamplight.

That morning, she was woken before 7:00 a.m. by Burley Elliott, an assistant who lived in a trailer on the property. They saw flames on the hills above them, and Curtin tried to find her eight dogs, which roamed during the day and returned home at night. Then she saw something that defied all logic and experience.

Looking across the road to her neighbor's property, just a few hundred feet away, she saw a bulging whirlwind of flame and smoke that was sucking debris from the ground, setting it alight, and rocketing it into the sky. It was a fire tornado as wide as a pickup truck, and it was moving slowly but inexorably toward her.

"It sounded like a freight train going around. It sounded. . . . It had this other sound too, more evil," she said. "Oh God, I can't even imagine what I could describe it as close to," she added. "There was nothing earthly about it." It was spinning and "throwing big pieces of wood on fire," three to four feet in length. This meant one thing to Curtin: *It's time to go, doesn't matter, leave everything and just go.*

Also at Curtin's home that day was a veteran named David Young. He went by the nickname Crystal Dave for the large, clear piece of quartz he wore around his neck, which his boyfriend had found locally and given to him before he died. Young told tales of his raucous adventures in San Francisco decades earlier, but now, with his blithe manner and flowing white beard, he was letting himself gracefully go to seed. He had a camper van and wasn't technically homeless, but he preferred to ramble from friend to friend and stay with them for long stretches. He thought the quartz "was invincible," Curtin said. "He thought the crystal could protect him from fire," from anything.

Curtin got in her cloth-topped Jeep, Elliott jumped on a quad bike, and Young took her Dodge Grand Caravan and her beloved Blue. "Just remember you have my favorite dog there," Curtin told Young. He left first. Curtin accelerated out with Elliott behind her.

Skirting the top of the lake, Curtin approached the intersection of Hoffman Road and Concow Road and decided she had to check on her rescues. The four draft horses were in a pasture beyond the intersection, on a dead-end turnoff.

A thunderhead of black smoke engulfed her, full of tiny embers that she couldn't help but inhale and that seared her mouth and gullet. Traveling at 30 miles per hour, it was almost impossible to see where she was going. Abruptly the airbags hit her in the face and chest, and every thought and sensation was knocked out of her.

As her mind coalesced the warm taste of iron bloomed in her mouth, and there was an odd, raw gap where all of her top front teeth had been. Every raggedy breath was accompanied by knife stabs in her chest, where she had broken seven of her ribs. The smoke cleared a little, revealing the grille of the truck she had collided with and the two men who got out. Bleary, bloody, she tried to open her door, but it was buckled from the crash.

Leave her there, she heard one of the men shout to Elliott. *It doesn't matter. We gotta go. The fire is too close.*

"Don't you leave, Burley," she said, "You come and you get me out of here."

Elliott had never intended to. He pulled Curtin by her arms out through the passenger door, and helped her on to the back of the quad, where she did her best to hold on to his back. The jangling, glassy agony of her broken ribs was beyond any pain she'd experienced before save that of childbirth. Elliott pelted south down Concow Road, and Curtin's hair momentarily caught fire from the embers, until they burst out into sunshine. For a moment, they had outpaced the fire. A fireman directed them to a pullout where they were to wait, because the blaze was about to come back over them. They did not know it yet, but Young had collided with a tree and been killed. Soon, Curtin was able to secure a ride to the hospital with two friends. "It's going to hit Paradise," she told them in the car. "The winds are too strong."

Hopkins and Edson, the two firefighters from Jarbo Gap, could go no farther. They had reached the 80-foot-long stream crossing at Hoffman Road. Depending on the time of the year, water flowing into the lake could rise higher than the bridge and come up to a passing car's fender, though now it was low. Four men had materialized from the flames near the intersection, one of them slightly on fire, his shirt smoldering, and jumped into the stream. Farther on, they told the firefighters, trees and power lines were blazing. Edson couldn't see a way forward and radioed that they were trapped.

Soon five or six cars had crowded onto the crossing with them, and others were backing up behind, laying on their horns. Hopkins grabbed a fire shelter—a protective blanket made of aluminum foil that firefighters only deploy in extreme circumstances, when a fire is about to blow over them or the heat is deadly—and ran along the line of cars. He found an elderly lady who seemed frozen with indecision or fear, pulled her out, covered both her and himself with the fire shelter, and led her back to the crossing. Gusts nearly ripped the blanket from his hands, and they were pelted by embers the size of charcoal briquettes. Shielding himself, he ran back for another older woman and her dog, supported a man who'd had a stroke and was unable to walk, and told three men who seemed like they could make it on their own to run for it. The beard of one of the men was smoldering.

By 7:45 a.m. or so, about twenty-five people were gathered on the crossing, some on the edge of the water holding up fire blankets to keep the burning embers off, others in up to their ankles and knees or deeper, splashing themselves to avoid being burned by the swirl of blistering projectiles. They were coughing, their throats felt like they were lined with sandpaper or glass, and their eyes were puce red from the smoke. Like a great bellows, the wind puffed up the flames dramatically, visibility was cut to 5 feet, and the firefighters shouted for everyone to get low in the water. Many had left their pets in the car. "In my normal mind I would have been screaming at that fireman to let me get to the car and the dog," said one of them, Karen Williamson. "But we were so fearful for our lives that I didn't care."

Engines and dozers were on their way, and as cars on the bridge caught fire Edson tried to call for a bucket drop of water, but the wind was too high. At last there was a glimmer of headlights through the poisonous murk, and a dozer materialized. They sardined the evacuees in several vehicles and headed to a nearby field that had been kept clear of vegetation so it could serve as a refuge in fires.

A former firefighter who lived in Concow named Scott Carlin had been doing his best to shut his mind off when he overheard a transmission from Jarbo Gap firefighters at the scene of the fire. Severe insomnia meant Carlin could go three or four days without sleeping, and that night, for the first time in years, he was forced to forgo his usual Ambien pill because he had been unable to fill his prescription. He lay oppressively awake on the living-room couch in a house west of the lake as his wife, Renee Carlin, a nurse at both an ER and a jail, and their two sons, Michael and Cody, dozed peacefully.

Carlin kept several of his old helmets on the wall, including the one he'd worn on a callout twenty years before, which almost became his last. He had been searching a burning mobile home for its occupant when a fireball exploded through it; he hit the floor, inches below the boiling maelstrom, and only got out because a colleague was groping for him through the doorway. Now he was a stay-at-home dad taking care of the property and of Cody, a sweet boy who had autism and needed ferrying to special classes and therapists. But it was hard for Carlin to shake the fire bug, which explained the radio scanners tuned to emergency frequencies on a desk just over his shoulder.

The message said the fire in Concow could be dangerous, but Carlin was cautious: was it really the case, or was the firefighter just green? When a second transmission confirmed the risk, Carlin roused his wife and sons, and they started throwing belongings in suitcases. Outside he tried to clear flammable brush away from the house with his little tractor, but it didn't have much effect. Another radio message said that a home nearby had caught fire, and Carlin ran five minutes down a dirt footpath through the ponderosas, to the lakeside home of his wife's parents, to tell them to get out. He jogged back to his own house, and then, for good measure, went back to his in-laws one more time to make sure they had left. Carlin joked that his mother-in-law, a retired teacher, did not operate on any schedule but her own. To his relief, they had gone.

But on the radio Carlin heard Edson's transmission about being trapped at the lake crossing. Having planned to escape this way, he was forced to change his course. Immediately he thought of the one place he knew would be free of fire.

Carrying their cases—with their cat in a cage and their two dogs on leashes—Renee and the kids hurried through the smoky woods, past her parents' house and toward the lakeshore. Cody had been afraid of dogs since a family pet had jumped on him and knocked him over when he was little. Renee told him he would just have to take one. Michael dashed to see if their aged neighbor, whom they only knew as Bruno, had left, and returned with Bruno's arm around his shoulder, half carrying, half dragging him down the hill. In no time at all they were engulfed. Flames were licking the edge of the lake, and the family edged out up to their knees in the muddy tule reeds. Bruno's foot got stuck and he fell on his back. Fragile at the best of times, he seemed to give up. *Just go*, he told Carlin. *Leave me alone.* When Carlin went to pull him up, Bruno refused to give him his arm. "I'm not doing that," Carlin said. "I'm not leaving you, I'm not gonna burn, so you'd better get your butt up."

Unexpectedly, the reeds began to catch fire, and they had to edge deeper into the lake. Renee and her sons were quickly up to their necks or treading water. Cody had the cat in a cage on his shoulder, and the mewling animal was half-submerged. It was early November in the mountains, and the water was so painfully cold it took Renee's breath away. She thought her heart might stop. Carlin, still in the reeds with Bruno, couldn't see her in the smoke.

Swim to the island, he shouted to them. By island he meant a peninsula sticking out from the opposite side, about 350 feet away. Cody stayed behind, but she and Michael kicked out in the frigid water, and in the half-light, Renee could see the shore around her wreathed in smoke, and orange flashes of flame. The husky she was holding got loose and swam back to Scott. The border collie, panicking, tried to climb on top of her and threatened to push her under, and she screamed.

Yet the wind was so strong that despite propelling themselves in one direction they were pushed in another, down-shore of where they had launched out. It was a lucky mishap—Michael found two rowboats, each with a paddle inside, tied with a cable to a stump. He tried to wrench the cable apart, but it wouldn't give, so he smashed the wooden post. Waterlogged and rotten, it splintered. Paddling themselves back to Carlin, they tried to coax

Bruno aboard, but he flatly refused to get in. He simply couldn't do it, he said. So with one hand Michael steadied a boat, and with the other he lifted Bruno onto the craft. Under the weight of four people and the dogs the boats were only two or three inches above the water, and Carlin told them to go without him. He would find them later.

They struck out, and this time they did end up on the peninsula. Soaking wet, they were freezing despite the occasional hot gust from across the lake, and they hid behind a rocky outcropping to try to keep out of the wind. Bruno's ability to make sense of the situation was going, and he seemed not to remember what was happening. Renee's legs were so cold that she couldn't walk. Michael went to find help, hollering in case someone could hear him. In the distance there was a response to his cry.

Relatives and friends of Pete Moak, who had been posting about the fire on Facebook earlier that morning, helped bring Renee and Bruno to the Moaks' home. Peggy Moak got Renee into a warm shower and dry clothes, and they wrapped Bruno in a blanket on a couch in the living room. Driving back around the lake, Michael found his father and the homes belonging to his parents and grandparents all relatively unscathed. They were among the lucky ones.

Firefighter Matt McKenzie was still in the Feather River canyon, patrolling the highway and trying to save what structures he could. And he'd been hearing radio traffic he couldn't quite make sense of. There was trouble in Concow, that he knew, but it seemed, incomprehensibly, as though the fire had jumped even farther, at speeds he thought almost impossible.

At 7:42 a.m., firefighters reported four spot fires across the lake. Within two minutes a caller at Drayer Drive, on the eastern edge of Paradise, reported fire in the canyon his home overlooked.

At 7:50, firefighters said they could see ten to fifteen large spot fires on the back side of Sawmill Peak. Officials requested six tankers, a lead plane, and six helicopters. At 7:56 a.m., dispatchers fielded a call about someone possibly trapped by the fire.

Two minutes later, Engine 81 in Paradise reported a spot fire at 1831 Apple View Way, a four-bedroom home near Paradise's ninety-seven-year-old Noble Orchards.

At 8:01 a.m., there was a fateful transmission from a patrolling aircraft:

"The fire is now in Paradise."

PART TWO
HELL

4

Daybreak

Before many people in Paradise knew there was a fire on the morning of November 8, it announced its presence with strange portents. Churning through the friable brush in the river canyons and in Concow, drawing power from the wind, it watercolored the atmosphere and flung little bits of itself far into the distance.

At a home in eastern Paradise, a man took his dog out into the yard for its morning exercise when he realized that the lawn was covered in burned leaves. It was so dark when another local rose at 8:00 a.m. that he thought the sun hadn't risen yet, though when he looked up he saw a red reflection in the murk. A senior went to a doctor's appointment and found that no one had shown up for work. But from somewhere in the distance he could hear metallic explosions. And at the Cypress Meadows nursing home, a manager was standing outside around 7:50 when a maintenance worker ran over holding a charred piece of bark: "This just came out of the sky. It was on fire."

In town, a forty-four-year-old homemaker called her daughters' high school to ask whether classes had been canceled owing to the plume she could see on the eastern horizon. They weren't going to

cancel classes for smoke, they told her. After all, fires were a regular part of life in Paradise. So she and the two teenagers hopped in
the car and drove to Paradise High. On the way they noticed that
the pines were coated in ash, as if there had been a blizzard.

In eastern Paradise, a former nurse was trying to motivate herself to get up when her ten-year-old daughter skipped into her
room in raptures over the sunrise. *It's so beautiful this morning*, she
told her mother. She asked for her mom's phone so she could
take a picture of it. The forty-seven-year-old, who struggled with
a rare disease that caused pain so debilitating it had ended her
career, forced herself out of bed to look at the dusky sky. When
she opened the door to let the family's barking dog out, she saw
that her garden was veiled with smog.

And a few miles away, when Patricia Smith walked to her
daughter's house along the path that separated it from her own
cottage, a black feather drifted past her nose. Smith had come
to America in 1983 from the British town of Boston and had
mostly lived in the Bay Area. But in 2017 she, her husband, and
her daughter's family had moved to a wooded, five-acre spread in
Paradise that they felt absurdly lucky to have been able to afford.
Smith's windows faced the forest—it was so splendidly secluded
you could walk around the cottage naked if you felt so inclined.

That morning, her daughter's alarm had not gone off, and they'd
all slept in late. At 8:30 a.m., Smith's granddaughter bounded
over to the cottage to wake her up and tell her about the fire. "Me
being English, I'm like, oh wait, I've got to make a cup of tea,"
Smith said. She drank half of it and had the cup in hand when the
feather floated past, inches from her face. Bending down to take a
closer look, she found it warm to the touch. It wasn't from a bird.
It was a charred wooden ember. She looked up. The horizon was
red but the entire rest of the sky was black—it was hypnotic and
beautiful and very, very wrong. She jogged the rest of the way to

her daughter's, spilling the remaining tea, and said: "We have to get out of here."

For the Paradise police, these warning signs were a call to action. Officer Rob Nichols had been roused at 4:00 a.m. by angry winds. Finally rising at 7:00 from a fitful sleep, he saw a rust-colored sky. While his wife helped their two young children get ready for school, Nichols called the station. A colleague told him about the blaze, but said Paradise would probably be fine.

A twenty-two-year-old rookie cop, Kyle Schukei, had arrived at the Paradise police station at 7:15 and was waiting for Nichols and a day of field training. He had been with the department for a month. It was quiet as he sat next to an officer catching up on paperwork. Soon he noticed a flurry of calls coming in, as residents asked about the thick black vapor outside their doors, and if a fire had sparked in town.

"As far as I know it's north of Concow," a calm dispatcher told one woman. "So far we're not in any danger." She rushed off the call as phones jangled in the background.

At 7:26 a.m., another caller remarked nervously: "Looking out our backyard, it looks pretty . . . it's kind of close."

As Nichols arrived at work, ash was falling from the sky and dusting the black-and-white police cruisers. He was walking toward the door in the station's back lot when Schukei and another officer ran out. They were getting calls for multiple spot fires around Paradise, and Nichols and the trainee were tasked with finding one just down the road from the Taco Bell in the middle of town.

"I'm driving today," Nichols said. It would be better to have someone who knew the roads well behind the wheel. Pulling out of the station, Nichols called his wife and told her to pack the car, take the kids, and flee.

As the officers neared the location of the call, Nichols spotted his wife's aunt and her granddaughter standing outside a home on Nunneley Road. He pulled over and told them, "There's a big fire, you need to get out. You need to evacuate."

Just a mile from the station, they found what all the auguries had been pointing toward. In awe they stared at an 80-foot ponderosa pine that towered over a central neighborhood of about thirty homes. There was a fire burning around the tree in a wide circle. Nichols, who had spent a season fighting fires in his youth, knew that the area must have been hit by a firebrand. The flames were working their way up the trunk and they were loud, making the dry bark and needles crackle like a fire on a camping trip. Embers were spewing into the air amid the surrounding homes and pines.

Nichols and Schukei raced from the cruiser and jumped over a 4-foot chain link fence to get a closer look. The extinguisher Schukei had to hand was no match for the engulfed tree. Nichols tried a garden hose, but nothing came out—the lot was vacant and likely had no water service.

Paradise S15, we need a fire truck, Nichols said over the radio, identifying himself by his radio call sign. *We got a pretty big fire here that's involving a large tree.*

The dispatcher responded moments later: *There's no resources available.*

Nichols was shocked. This could only mean that every firefighter in town was already occupied somewhere else, and that what was happening in front of him was likely happening all over Paradise. His gut told him to get people out.

The pair ran from house to house, banging on dozens of doors to alert people that they had to go. Some were still asleep. At half of the homes no one answered. It took them twenty minutes, and when they were done they got another call for a spot fire near the school district office on Clark Road.

The streets were already crowded as they drove the two miles north. The blaze was still small by the time they arrived, and Schukei tried to put it out with the extinguisher. Nichols scanned the surroundings, vacant lots and the Kmart parking lot across the road. His heart sank. Another massive ponderosa, this one 50 feet in height, was swathed in flames from base to crown. Fierce gusts of wind scattered burning pine needles all around it. Nichols was not a man prone to melodrama, but at this moment he felt defeated.

We're done, he thought to himself.

Paradise had prepared for an emergency evacuation, but never for one of this magnitude. The first alert, sent via robocalls, texts, and emails, went out at 7:57 a.m. to residents in eastern Paradise along the edge of the canyon, in three of the town's fourteen designated emergency zones. "Due to a fire in the area, an evacuation order has been issued for all of Pentz Road in Paradise east to Highway 70," it said. Minutes later, at 8:02 a.m., officials ordered the mandatory evacuation of the entire town, and began triggering more warnings.

Yet less than half of the 27,000 residents of Paradise had signed up to receive alerts in the first place, a problem that plagues opt-in systems. In the eastern neighborhoods first hit by fire, more than half of the 4,272 alert calls failed, because phones had been disconnected, numbers had changed, or cell towers didn't work. They may have burned or been overloaded by the flood of alerts put through the system. In the western portion of town, three zones did not receive orders to go.

In total, only 6,200 people in Paradise received official warnings about the approaching wildfire. Most residents learned about it from neighbors, passing sirens, lines of cars in the streets, and spot fires in their yards. One woman got an alert to evacuate as she stood outside her childhood home watching it burn.

Emergency dispatchers were fielding calls from increasingly panicked residents. A woman named Ann called 911 because she didn't have a car in which to evacuate. She couldn't leave with neighbors, she said, because they wouldn't take her dogs.

"Ma'am, you need to save your own life. I understand your dogs are precious to you, but you need to save your own life," the dispatcher told her. "I'll pass it on to the police department that you're having a hard time but I can't guarantee that we can get anyone there to help you. So you probably should go with the neighbors, OK?"

The woman began to sob.

"Leave the dogs inside, leave them water and food and go. Do you understand me, Ann? This is a serious fire, I need you to leave the dogs with some water, close all the windows and the doors and get in the neighbors' car. . . . I need you to do that. . . . You need to do that right now, OK?"

A minute later, a man called for information. "There's fire everywhere, OK?" the dispatcher said. "There's fire in Paradise. We have fire all over Paradise. If you feel unsafe, you need to evacuate."

"I'm not sure. What do you . . ." the man asked, trailing off.

"If it was me I would get out of there," the dispatcher said.

Another woman said she and others were surrounded by fire and trapped in their vehicles behind a fence at Ponderosa Elementary School.

"There's no one to come help you right now," the dispatcher told her. "If you need to drive through a fence, then do it."

Other emergency procedures were quickly and rudely tested. City manager Lauren Gill, Paradise's top administrative official, woke at about 7:00 a.m. and flicked on the lights to see if the electricity was on. Looking out her window, she noticed that the sky was a gauzy, yellow-brown color, and when she opened her door she saw a nasty plume. Checking her phone, she found a missed call from the Paradise fire chief, and ringing him back, she learned about the fire in Pulga. He told her it was still far away, but Gill knew she would have to keep the town informed, and she hurried to the wooden town hall. As soon as Gill walked in, she learned that evacuations had already been called. Paradise had prepped for disaster for years, but the speed of this one was inconceivable. "It's almost like you wake up in the middle of a nightmare," she said. "You think you're going to wake up to a day, and you wake up and you're in the middle of this drama, this emergency, this nightmare that's occurring around you." She told the staff gathered there: *We're having an EOC.*

Paradise's Emergency Operations Center is activated in the event of a disaster and comprises officials who coordinate the response. But during the Camp Fire, the officials went out to reconnoiter the streets and did not return because they were pulled into assisting with the evacuation and saving lives. Gill's phone was flooded with calls, and her staff soon told her that the power was out, as was cell service and the Internet, and that the town hall was in the path of the flames. Gill ordered the EOC to transfer down the hill, to Chico. She wanted to be the last one to evacuate, and she asked a retired fire chief stationed with her what would happen if the two of them stayed behind. He told her they could lay on the ground in the parking lot. Gill then asked if he had any protective fire blankets. He did not.

"OK," she told him, "then we'd better go."

At 6:00 a.m., Cal Fire bulldozer operator Joe Kennedy arrived for work at his fire station in Nevada City, a 75-mile drive southeast of Paradise, in the same wooded foothills. Soon wildfire alerts echoed through the forest clearing where the fire station was located, and pinged his beeper and cellphone: *Dozer 2342, out of unit response, vegetation fire to Butte.* "Out of unit" meant that Kennedy was being dispatched beyond his home region—in this case to Butte County.

Firefighters used to talk of particularly severe fires as "career fires," meaning they'd see one of them in their professional lives. Kennedy was only thirty-six, and had joined Cal Fire in 2014, but he'd already had four such fires. In the 2016 season alone, he'd been deployed as far afield as the Mexican border and the Oregon border. During one numbing stretch, Kennedy had been up for four days straight when he and another dozer operator both fell asleep at their wheels and collided with each other.

As is not uncommon at Cal Fire, the stress of the grueling work built up inside him, and it was difficult for him to convey to others how deeply he was affected by it. "It's hard to fight and be in that mindset and then call home to your wife and kid and be soft," said Kennedy, a 6'4" man who looked as imposing and unshakable as a steel beam. Yet he also let his four-year-old daughter paint his nails and he took part in her tea parties.

Kennedy came to firefighting late. Raised by a single mom, he never enjoyed academic subjects at school and always wanted to be outside, engaging with the world physically. He got into fights, not because he liked to cause pain but because of the rush of it and the strategizing it required to win. His first job after graduating from high school was as a welder, and he progressed into operating heavy equipment for a pipeline construction job. Heavy machinery connotes crude, brute force, and one of the nicknames given to him by his colleagues was Bamm-Bamm, after the club-wielding baby in the *Flintstones*. "Because I'm good at breaking stuff," he said.

But Kennedy found the job required finesse and rhythm. He had to figure out how to balance about 40,000 pounds of machinery, to feel how the machine under him was responding to the terrain and the loads he picked up, and to progress smoothly and methodically through space. It seemed to Kennedy like dancing or playing the piano.

He switched to firefighting because he felt a call to be of service. And a terrible fall on a hunting trip a few years earlier, when he had a lung embolism and had to be put in a coma for twenty days, made the prospect of solid, state-provided health insurance appealing. Cal Fire was not what he expected. His crew was mostly male, and it surprised him to find the station immaculately clean. It was vacuumed or swept daily, and the engines were kept pristine and gleaming. Firefighters prepared plated meals for each other at precise times. All of this suited Kennedy, a two-showers-a-day kind of guy. It was good for morale and camaraderie, and it was only right to treat tax-dollar purchases with respect.

When the "out of unit" call came through, Kennedy jumped into the cab of the eighteen-wheeler that pulled his dozer and, to the clamor of sirens and flashing lights, covered the journey to Paradise in about an hour. The roads were empty until he reached the town and traffic started to congeal around him. Often bulldozers operate outside a fire. Their job is to create firebreaks by mowing strips of land around the perimeter, tearing out trees and bushes and the top layer of dirt, down to the noncombustible mineral soil. The "fire line," as it is called, should be at least 1.5 times as wide as the vegetation fueling the fire is high. When wildland firefighters give percentage figures for how much a fire is "contained," this is what they are referring to: the proportion of the fire that has been boxed in with fire lines by bulldozers and hand crews.

Kennedy's destination was unusual. He wasn't headed for the perimeter. He had been assigned to the Feather River hospital, on the eastern border of Paradise, where he was to bolster the firefighting division already at work nearby. The hospital was a crucial location in those first hours, right in the blaze's path as it steamrolled in from the east.

Meanwhile, hurried escapes were proceeding at homes and institutions all over town. At Cypress Meadows nursing home, Sheila Craft, the admissions and marketing director, ordered an evacuation after spotting fire racing across the canyon. She made urgent calls to facilities in the valley to find places for residents: *We have a fire up here and we have a crisis situation. How many beds do you have available?* The team piled patients, some recovering from hip-replacement surgeries and others with dementia, into staff members' cars and police vans that had rushed over from the sheriff's office. At Apple Tree Village senior mobile home park, maintenance man Stephen Murray drove round with his horn blaring. He banged on residents' doors, and if they didn't answer he kicked the door down.

At Paradise High School, sisters Arissa and Arianne Harvey, sixteen and seventeen years old, were dropped off by their mom at 7:20 a.m. They had breakfast in the cafeteria—Arissa, who studied Spanish and Korean in her spare time and was developing a fondness for Russian novels, was excited for the cinnamon rolls, "so soft and squishy and very sweet and warm." Looking out the large windows, Arissa remarked to her sister, *It's snowing.* Except, of course, that it rarely snowed in Paradise. The sun, invisible, cast an entrancing glow.

Arissa's first class, history, was at 8:00 a.m., and before it began she headed to a neighboring building to pick up some sheet music from the choir director. The hallways, normally crammed with people, were unexpectedly empty, and as she hurried along a covered outdoor walkway the sky suddenly filled with birds flocking out of town. "I've never seen so many birds fly together," she said. "Just a constant stream of birds for a minute, a minute and a half." There was an endless, barking *caw* of crows calling to one another.

The bell rang for the start of school, but it seemed that only ten seconds later there was an announcement over the loudspeaker telling students to go home, and to get a ride with a teacher if they couldn't leave with anyone else. Arissa found her sister and they called their mom, who said: *I'm already heading back.*

Arriving at Paradise elementary school, a teacher named Lynn Pitman had seen smoke and thought little of it. When it became so stifling that the children were sent in from the playground, she began to worry. Sitting at her desk to call parents, "I was shaking pretty bad." Once only a few kids were left, she gathered with the other teachers and their remaining students in a classroom upstairs.

Outside it was now growing dark, as if an eerie twilight were falling, even though it was still morning. "Looking at the kids' faces I thought, *I gotta do something with them*," Pitman said. Forcing cheeriness, she gave them snacks and began playing games, until police came through to check the building and told them to move. Buses were waiting next door, they said. "Once I got outside, the smoke, the darkness—it looked like midnight," Pitman said. "Then I got really scared."

When smoke rose from the canyon by the Feather River hospital, the wildfire-hardened medical staff weren't worried. At the sprawling facility, patients streamed in just like any other morning, complaining of chest pain or getting ready for scheduled surgery. In the emergency room, a nurse named Chelsea West sat drinking coffee and readying herself for the fourteen-hour day. During her twenty-minute commute from Chico to Paradise, she had been listening to the news—more updates from Tuesday's midterm elections, and the grim body count of a shooting in a Southern California bar—when fire trucks sped by with their lights and sirens. It was her last hospital shift of the week, and tomorrow she would be in a studio in Chico teaching vinyasa yoga. It was a change of pace she looked forward to each week. West thought the relentless nature of ER work had made some at the hospital cynical, and feared it could happen to her.

On the manicured grounds, the hospital's intensive-care manager, Allyn Pierce, was watching smoke infuse the sky as ash rained down over him. Pierce, a forty-two-year-old, shaggy-haired mechanical engineer-turned-nurse, had been at Feather River for about fourteen years. The son of a doctor and a nurse, he had virtually grown up in a hospital. But it was only shortly after he moved to his wife's hometown of Paradise, and she'd cut her finger so badly on a can it required a trip to the ER, that he decided to go into healthcare. He was reminded of how at home he felt in the medical world, and he had a stomach for dealing with gore and chaos. At Feather River he worked alongside colleagues he admired and he enjoyed treating two-month-olds one day and World War II veterans the next. Pierce had been offered the ICU manager position four years earlier and had settled into it; he thought it was the job he would retire doing. Paradise was sometimes too quiet compared to his native Southern California, but it was a good place to raise his two kids—and affordable.

The ICU deals with the sickest patients—those who can't breathe on their own, for instance, or who have head injuries that require constant monitoring—and by about 7:45 a.m. Pierce had directed his team to ready the wheelchairs and gurneys.

At an emergency meeting Pierce and other hospital leaders discussed a possible evacuation. The hospital's chief financial officer, who was in charge that day according to the emergency protocol, had been in communication with Cal Fire. "We knew where things were headed," Pierce said.

The meeting started at 8:00 a.m. and took just nine minutes from start to finish, Pierce noted. In the final moments of the meeting, the CFO ordered the evacuation. It was a "code black"—*get everyone out*. Pierce and the other team leaders began alerting their staff while embers showered the hospital.

They would have to evacuate sixty-seven patients—some very ill and on life support, one anesthetized for a gallbladder surgery, and at least one more on the operating table for a caesarean section—and judging by the smoke, they wouldn't have long to do it. It was a fitting day for the motto under which Pierce operated the intensive care unit, particularly when it was busy or there weren't enough staff: *It will work out because it has to.*

In the hospital's maternity ward, Rachelle Sanders was already awake when a staffer came into her room and said she might be evacuating. Sanders's son, her third child, had been born by caesarean at 8:30 p.m. the night before, two weeks early but a healthy six pounds. Lincoln, a fuzzy-haired blond boy, lay in a bassinet next to her. Sanders was still immobile, tethered to an IV inserted into her arm and a catheter. She couldn't get out of bed, let alone walk to the bathroom. But, not overly concerned, she encouraged her husband, Chris, to leave to check on his mother, who was visiting from out of town and staying at their home a few minutes away.

A nurse came by to fill out Sanders's morning chart, listing her temperature, blood pressure, and pain level. After a quick chat, she stepped out, but burst back in a few moments later. *We're evacuating now*, she told Sanders. The nurse said it with such urgency that Sanders felt like she was being asked to pack up then and there, but it had been less than twelve hours since she delivered Lincoln. Sanders thought to herself, *What do they want me to do?*

Two orderlies hurried in and lifted Sanders out of bed and into a wheelchair, gently placing Lincoln on a pillow in her lap.

They stationed her at the front of a long line of patients winding through the hospital to the ambulance bay out back. But the hospital didn't keep emergency vehicles on hand—ambulances were based at locations across the county and had to be called in—so staff queued up in the bay in their own cars. Sanders was bundled into a four-door Nissan driven by a man she took to be the hospital's maintenance engineer. Before they left, hospital staff handed her a mask, hung her IV fluids on the rearview mirror, and placed the catheter bag on the floor. Lincoln lay on the pillow in her lap, sleeping peacefully.

They were among the first to leave the hospital. Sanders couldn't get hold of Chris, but she spotted his car ahead of theirs and told the engineer to follow him. She warned him that she would have to undress to feed the baby. "Don't worry," the engineer told her, "I've been married a long time. You just do what you have to do and I'll focus on driving."

By 9:00 a.m., about an hour after the evacuation had begun, the hospital was mostly clear save for staff, and Allyn Pierce was directing his team to get out. It looked like nighttime, and the light of the departing ambulances and the glow of the fire in the clouds cast a purple hue over the buildings. Nurses and doctors were running to the parking lot. "Who needs a ride? Anybody?" someone shouted. It would be the last time several of them ever saw each other again.

As many of the tens of thousands of inhabitants of Paradise, and of Magalia higher up in the hills, all tried to leave at once, the result was predictable. Cars, RVs, and trailers soon created a massive traffic jam as the fire outpaced them and trees and homes on either side of the road went up like torches. As they inched along, some vehicles ran out of gas or caught fire themselves, blocking the way and making the backup worse.

Paradise officials had never contemplated a fire as terrible as this. "We always thought a fire, even if it was a fast-moving fire, would burn into a zone or zones," basically just a handful of them, said Jim Broshears, the former fire chief who had helped craft the emergency procedures. "Not that it would jump over three zones into the fourth zone and then sideways and two zones up, and eventually burn in every zone." Far from moving logically, the fire blew past some neighborhoods, sparing them initially, only to eat its way through them several hours later; spot fires that erupted to the west of town burned backwards, in the direction of the ignition point.

Dozens of first responders were stymied by the snarl of vehicles. Even the highest-ranking, including the Paradise police chief, the county sheriff, and a member of the town council, found themselves with the humble task of directing cars through the clogged streets. At the corner of Skyway and Elliot, twenty-four-year-old Paradise police officer Kassidy Honea, the only woman on the patrol force, dodged drivers who thought they could get down the hill quicker if they rammed through orange cones, and told people frozen with fear: *Keep going or get out of the road.* Working with her was a colleague who heard reports of deputies trapped up north and told her he was going to help. "I'm so proud of you," he said as he wrapped his arms around her. It was her father, Butte County sheriff Kory Honea.

Those departing the hospital soon ran into trouble.

A second ambulance carrying another woman who had just given birth broke down in the thick smoke and caught fire. Another ambulance carrying patients stopped behind it. A pediatrician placed the immobilized woman on a backboard, and David Hawks, the Paradise fire chief, arrived and shuttled her and the other patients and hospital staff toward a tan stucco home on Chloe Court—the only one in sight on the street that had not caught fire. A paramedic slipped through a dog door and unlocked the garage, shepherding the group of a dozen inside as fire rained down over the street. With the patients secure, Hawks directed the hospital workers and medics to try to protect the house from flames by clearing brush, hosing down the roof and removing pine needles from gutters. They sheltered in the home for two hours before a sheriff's van was able to rescue them.

Some of the last staff to leave Adventist Health Feather River split up into groups.

ICU manager Pierce and two colleagues headed down Pentz, a two-lane road that was already on fire, in his massive Toyota Tundra truck, which had been packed for a visit to Joshua Tree with his daughter that weekend. To keep things light, he put on the only upbeat music he had: the soundtrack to the superhero film *Deadpool 2*. A song by Celine Dion called "Ashes" came on. It was a bit too on the nose, so he skipped it. The group was quickly diverted down Pearson Road. All around them, enormous trees were funneling flames into the sky. Stuck in traffic behind a gold Toyota, Pierce took photos and recorded videos of the scene, and himself. Wearing brown, thick-framed glasses and a mask, he joked: "I think my camping trip is canceled."

The traffic was so bad they barely moved. A Volkswagen Beetle passed Pierce's truck and caught fire. Its driver jumped out and ran. The houses around them burned, and Pierce could hear propane tanks exploding in the distance, one after another. Cars, some abandoned, completely blocked the road. In the fire truck next to him, firefighters began covering the windows of their truck with heat-reflective blankets—a last resort. Pierce's colleagues left and sought refuge in the cab. But Pierce stayed: he hoped the road might somehow clear, and he didn't want to trap anyone else behind him by leaving his own vehicle.

Although the air conditioner was running full blast, the windows were hot enough to burn skin. The heat started to char the white truck like a marshmallow and to melt the wheel wells. The firefighters requested an air drop. It was starting to seem to Pierce as if he might not make it. He recorded a video for his family and friends, and buried it deep in his center console, hoping that it would survive if he did not. "In case this doesn't work out, I tried everything I could. I'm doing everything I can to get out of this. I love you all," Pierce said. He envisioned dying smothered in burning plastic and metal.

Listening to an acoustic version of A-ha's "Take on Me," he put his hands on the wheel and sang along to the chorus, trying to blot out the whirls of fire outside the window.

At about the same time Pierce left the hospital, Chelsea West and three other nurses decided there was no way they would make it out by car. Traffic had backed up into the hospital parking lot, the power had cut out, and small fires glittered all over the campus. Wearing surgical masks, the four chose the simplest option: they walked away from Feather River out onto Pentz Road.

Flames danced in the brush as the wind showered them with glowing pine needles. A dark pall churned above their heads, and the smoke made it difficult to breathe even with their noses and mouths protected. Scared but determined to make it out of Paradise, West and her colleagues had trudged past burning homes and trees for only a few minutes when they chanced upon a fire truck and stopped to see if they could get a ride out. A firefighter asked a twenty-three-year-old sheriff's deputy named Aaron Parmley, who was sitting in his cruiser, to see if he could evacuate them.

A second-generation cop from the Central Valley, Parmley typically worked just a few miles up the road out of the Magalia substation. Quieter than much of the county, the community was also affordable enough for him to buy a home on a third of an acre, with room for him and his orange tabbies, Lily and Sissy, to spread out. Good fishing and camping weren't far, and he was close enough to work that when a colleague texted him about the fire, it took the goateed deputy five minutes to get to Paradise.

Parmley had just finished evacuating a retirement community behind the hospital. Black smoke billowing in through the vents of his car had forced the asthmatic deputy to turn the air conditioner off, and he was wondering if he might pass out. But West and the other nurses gratefully got in with him, and Parmley headed south on Pentz Road, toward the valley.

Soon he discovered that their escape route was blocked by downed power lines. Parmley veered right looking for another way out, a decision he immediately regretted. Smoke was especially thick on Pearson Road, reducing visibility to only 20 feet, and fires burned on both sides of them. Embers clattered onto the hood of the car. As they crept down the street, Parmley stopped to assist a California Highway Patrol officer whose vehicle was inoperable after a carful of evacuees crashed into him head-on. They were huddled together and one of the women was barefoot, Parmley noticed, and was treading on hot embers. The scene left Parmley with a sense of deep foreboding, and he helped the stranded group flag down rides from the handful of motorists passing by. When he and the highway patrolman hopped back into his cruiser, something seemed wrong with it. He'd left it on, but the engine had switched itself off.

The car was quiet as Parmley tried to start it, and the silence punctuated by the clicking of the ignition made West's heart sink. There was no one else on the road by this point, and outside the flames were growing brighter and taller. *Not like this*, she thought. *I don't want to burn.* For a few minutes, they waited while the deputy tried to put out a call for help. The radio seemed to be working, but dispatchers couldn't hear him. A nurse pulled out her cellphone and they took turns making their goodbye calls.

"I love you. I don't think I'm getting out," West told her boyfriend of seven years, tears wetting her smoke-stained mask. He cried, too, and told West he loved her. Their conversation lasted only a few moments, so that everyone would have a chance to try to contact a loved one.

The deputy put out one last call for help: "My car is disabled. We'll be walking on Pearson toward Pentz Road."

This call seemed to go through.

"Copy," the dispatcher replied.

Still, Parmley thought: *I'm done for.*

West was determined to get out. The nurse had the steely expression of a soldier. Feeling stifled, she took off the scrub overshirt she wore, dropped her purse, and set off back the way they came. Parmley and the others went too.

Struggling to breathe as embers and thick smoke swirled in the air, they trudged through the dark hoping for some way out. At 9:51 a.m. Parmley turned on his body camera to record what he assumed would be his final moments. The sky was dark as they walked but the ground was aglow. The wind swept embers back and forth over the roadway like amber snowflakes. "It's bad," Parmley mumbled over his radio.

They were all tired. Some of the nurses stopped to try to catch a breath of clean air that wouldn't come. At some point West rolled her ankle and broke her foot. The group had dispersed over about 100 feet. Parmley was at the back with the patrolman and the nurses, running ahead, had nearly disappeared into the smoke. *Keep going!* someone yelled.

On the body-camera footage, sounds of heavy breathing mixed with the crackle of fire and the mumble of voices over Parmley's radio. The scene looked like the surface of the sun, or the molten core of a volcano. Hot coals danced about them and the burning hillside appeared subsumed by lava.

"Are they coming for us?" asked a nurse. It seemed like they were completely alone.

Pulling up at the hospital with his bulldozer, Joe Kennedy didn't know where to park, but the question was decided for him. The truck's wheels sank and got stuck in a well-watered patch of grass. He encountered a reporter who seemed frightened and asked if she could stay with Kennedy. He told her to hurry back down the road and get out—he was heading in the opposite direction. The brief exchange made him notice that his own adrenaline had ratcheted up. Paradise was submerged in an acrid cloud that pulsed red from within.

Kennedy unloaded his dozer and radioed for directions. No one answered. Without explicit instructions, he went to work trying to save lives and structures. He came across volunteer firefighters attempting to protect a house and cut some line for them, pointing out something they had missed: the fire was coming at them from behind. Continuing on, he used a tactic he called "fire-front following." Many structures can survive the sweep of the main flame front, but they are still at risk from the small fires and thousands of embers it leaves in its wake. Kennedy proceeded to save five or six more buildings by mopping up these little blazes until a staticky call hissed over the radio: it was an engine in his unit, trapped and asking for an air drop. He knew this would be impossible because of the howling wind and smoke, and at his top speed of 7 mph, he headed west into the blackness, in the direction of Pearson Road, which traverses the lower half of Paradise from east to west. The metal band Pantera screeched from his stereo, pumping him up like during a workout.

Kennedy was proud of the fact that he ran toward fire rather than away from it. He was one of 2,289 firefighting personnel who raced to the Ridge that day, on equipment including 303 engines, 11 helicopters, and 24 bulldozers. Fire crews mobilized across Northern California—from towns such as Redding and Auburn—and beyond. Anticipating the call, an incident meteorologist in Reno, Nevada, packed his bags with computers and camping gear. Tim Chavez, a thirty-five-year veteran with Cal Fire, received the long-anticipated notification in his Southern California home. The Los Angeles–area battalion chief was on one of the agency's incident management teams, and he had predicted fire when he saw the dry and windy weather forecast for that week. Chavez had fought fires across most of the state, but he thought of Northern California blazes as wildcards. "They just go on and on and they don't stop," he said. "You get the helpless feeling like, *Is this ever gonna end?*"

Just before 7:00 a.m., Shem Hawkins, a Cal Fire battalion chief in charge of the Chico Air Attack Base, had received a call from one of his tanker pilots who could see a rapidly building smoke column in the canyon through which the North Fork Feather River flowed. After conferring with other commanders, Hawkins advised his team to convene at the base, located at the Chico Municipal Airport. Firefighting planes are unable to extinguish fires by themselves— their role is to box them in with a retardant slurry composed of water, ammonium phosphate (used in fertilizer), and gum (to help it stick to vegetation), enabling ground crews to squelch them. The aircraft release the red or pink goo flying at approximately 150 mph, though only in certain conditions. If the wind is too strong it disperses the retardant and renders it ineffective. Helicopters are better suited to poor conditions because they can hover at lower altitudes and deposit water with greater accuracy.

At about 7:40 a.m., the first firefighting plane, a converted Grumman antisubmarine aircraft, took off and attempted to drop retardant on a communications tower east of Paradise that first responders relied on. But when the pilot cut the engines to begin his glide-and-drop sequence, the gale was so strong that instead of descending the plane was pushed higher, and turbulence made it difficult to control. Fixed-wing retardant drops were out, Hawkins realized, and the Grumman returned to base.

Hawkins himself took to the air in the jump seat of a small, double-propeller reconnaissance plane called a Bronco. His first fire job had been as a reserve firefighter in Magalia, and he went on to work as a paramedic. Now his assignment was to coordinate aircraft and ground resources. Down below "you have a very narrow optic of what is occurring," he said. "From the air, you're witnessing the whole thing." At 7,000 feet, he was staggered by the smoke column, a measure of the fire's intensity. "Usually it takes a little bit of time for that column to develop," he said. "That morning what I saw was the fastest growing column I've ever seen in my life." It had expanded as much in forty-five minutes as he would have expected in a few hours. Spot fires were metastasizing in Paradise. He requested six additional air tankers and six helicopters, as well as an "aerial supervision module"— another aircraft that would help coordinate the fleet. He directed two California Highway Patrol helicopters to a spot where they could land in Concow and pick up burn victims, and he ordered other helicopters to drop water along the evacuation routes and at the hospital.

This fire was especially fraught for Hawkins. He had grown up in Paradise, and down below, his own sister-in-law and her children were trying to escape. "You disassociate yourself somewhat from it—at that point you're focused on doing the job," he said. "It was afterwards: I understood that community and knew who was there. I knew it was a very geriatric population. I knew they would be moving slowly and couldn't outpace that fire."

On the ground, everything was burning bushes and homes. Kennedy was heading toward Pearson Road to help the trapped fire truck. He saw a light flashing in the darkness, and thought it was the spark of a downed power line, until it flickered again. He was astonished to see the silhouettes of several people emerge from the crimson billows.

Deputy Parmley, the patrolman, Chelsea West, and the other nurses had heard a loud crack, and West worried that a tree was coming down on top of them. Instead they saw lights. The bull-dozer that appeared in the distance looked like a train, its beams cutting through the thick smoke. West and the others franti-cally waved their arms. Parmley grabbed the light off his belt and flashed it again and again, hoping the driver might spot them. When the machine came to a stop, he rushed over.

"Can we get in?" Parmley yelled.

"I have space for one, two," Kennedy replied.

Parmley had been so happy to see the dozer, and was crest-fallen when it was clear there wasn't room for everyone. Kenne-dy's heart sank briefly as well—he had been diverted from saving the engine in distress. Two of the nurses got in. But another fire truck that was following behind him sounded its air horn, and all the nurses and Parmley piled into that instead.

Kennedy made it about another quarter mile before he was blocked by dozens of seemingly abandoned vehicles. The fire was moving through them like it would through vegetation, and Kennedy decided to treat them as such: he would separate whatever was on fire from whatever was not. The first car was a Prius, combusting with such intensity that as he engaged it with his blade the bulldozer windows cracked and the external lights began to melt. Kennedy inhaled one or two searing breaths and thought for a moment that this had been a fatal decision. But the temperature fell, and he continued, though now he could only see out of a small, angled window on his left. In one abandoned car he caught sight of a baby seat. Others were not abandoned at all. He saw what appeared to be the burned bodies of people who had not been able to get out in time. Worse, he had to shove them and their cars off the road.

Finally he reached the trapped engine. It was so hot around the fire truck that the paint was peeling off the outside. Now it could move. And, just next to it in his Toyota, so could Pierce, the ICU manager who'd nearly lost hope. Pierce realized he would see his wife and children again. *I can do this*, he thought. He could have fled, but instead he drove back in the direction of the hospital. There would be people who needed his help.

By mid-morning, the fire was a different kind of blaze from the one that had roared into existence only four hours before. Instead of a wildfire out in the countryside, it had become an urban fire spreading from building to building. The heat of burning structures ignited vegetation rather than the other way around. Showers of embers found their way into yet more buildings by means of ventilation ducts, cracks, and gutters full of leaves. The Safeway, the Skyway Villa RV park, the Paradise Adventist church, the mayor's house, the police chief's house: all went up. In a cruel twist, many buildings in this lower-income community were more susceptible precisely because they were older and closer together. "It went from home to home," said battalion chief Ken Lowe. "We couldn't catch it."

Fire surrounded the Feather River hospital and outbuildings were burning, but the main hospital still stood. The police station survived thanks to volunteers who hosed the building down, although fire destroyed the neighborhoods immediately north of it. Paradise Elementary School burned and the play structures melted.

At 9:43 a.m. a commander made an announcement over the radio that reflected a new mission objective. No longer was the goal to restrict the fire to a specific area: "We're just working on evacuations. That is our only priority right now. There is no fire defense going on." The chaos meant that firefighters, rather than following precise orders, were given greater authority to make their own decisions, making the Camp Fire one of those atypical blazes when the response is based on the principle of "leader's intent."

By 10:00 a.m. the fire had consumed 6,000 acres, more than nine square miles. The main body of the fire was working its way toward Clark Road, the thoroughfare that cuts north-south through the center of town, while spot fires burned throughout Paradise. Within minutes, firefighters would report that flames were well established in Butte Creek Canyon, on the western side of Paradise. Satellite imagery showed masses of flames surrounding the town from two sides.

And the death toll was starting to mount.

5

Stay or Go

In the face of natural disaster, there are few options besides *run* or *hide*. No one tries to divert a hurricane, stifle a tornado, or tame an earthquake. But Americans claim dominion over fire. The firefighting response to wildland blazes is as muscular as that of an army in wartime. The infamous urban conflagrations—San Francisco's immolation after the Great Earthquake of 1906, the Chicago Fire of 1871—are a thing of the past owing to the evolution of building codes that mandate the provision of sprinkler systems and emergency exits. With success comes certitude. It is hard even to imagine that a fire could invade a modern American community, rampaging from a KFC to a Dollar General, from a fire station to a suburban subdivision. Individual structures still burn, but the city as a whole has become inviolate.

Yet as towns sprawl in the flammable, warming countryside of the American West, it seems premature to write wildfire off. "We're trapped by the myths of our own success," Jerry Williams, the retired national director of fire and aviation management at the US Forest Service, once said. "Sometimes I think this is almost a religious issue, that we can somehow dominate it."

On the Paradise Ridge on the morning of November 8,

amid great bravery and tragedy, these myths were to be brutally dispelled.

By 9:00 a.m., the two-lane road that curved north past the Sedwick home in Magalia, just north of Paradise, was already blocked
and at a standstill with cars trying to evacuate. *Don't go that way*,
John Sedwick warned his daughter, Skye, as he stood watching
the flames tear through the canyon in the distance. *People are crazy
down there. Go through the woods.*

The "woods" meant Glover Lane, a forested road that cut
through Sedwick's property and wound downhill past an abandoned trailer and clandestine pot gardens for about half a mile
before meeting Old Skyway next to the Magalia Community
Church. Largely unpaved and overgrown, it hadn't been much of
a thoroughfare in a few decades, and vehicles could no longer tra-
verse it in its entirety. Though Sedwick's address was on Glover
Lane, a neighbor had built a home in front of his property and cut
off his access to the road. Still, he and Skye often strolled along
it—she would sometimes collect pine cones to use as kindling in
the woodstove.

Skye dressed in gray that morning—her sweater, frayed skinny
jeans, and Chuck Taylors were all the same gunmetal color. It
seemed the appropriate shade, Skye thought to herself as she
quickly got ready, applying the mascara, eyeshadow, and foundation she wore every day. Carrying her purse and a bag on each
shoulder, tightly gripping the leash around the neck of her dog,
Jude, she headed down Glover to meet her boyfriend, Merritt. It
was still early and should have been cool, but it was surprisingly
warm underneath the massive ponderosas and Douglas firs.

Skye had just passed the abandoned trailer, three minutes from
home, when she spotted flames. Leaves, pine needles, and the
blackberry bushes from which she often picked fruit had ignited.

The fire was about a foot high and being fanned by the wind, which sent burning leaves swirling through the air. Embers dropped from above.

Skye started to get anxious, and she thought about warning Sedwick but decided he knew what he was doing. As she maneuvered herself down the slope she almost slipped. Jude whined, his ears flattened, and he had a bowel movement. "It's OK," she told him, taking a deep breath. "We got this."

When she finally came out next to the church, Skye had gotten her nerves under control, but a feeling of unease lingered. In the distance, she could hear a *bang, bang, bang* of detonating propane tanks, and a house was smoking across the street, next to the old Magalia Cemetery.

The well-watered lawns were surrounded by imposing Douglas firs and mosaicked with the graves of some of the area's earliest Gold Rush–era settlers. Skye and Sedwick had visited the grave of Sedwick's first wife there countless times. Each year, on his wife's birthday or their wedding anniversary, Sedwick or Skye would pay their respects with flowers.

Walking on the narrow gravel path alongside the two-lane road, Skye headed south toward the Fastrip gas station, a mile away in Paradise, where she'd arranged to meet Merritt, but within a few minutes, drivers on Skyway began stopping for her. *You can't leave walking—things are exploding down there,* one said. Skye and Jude eventually accepted a ride with strangers, a woman and her mother, headed to Chico.

As they fled Magalia, traffic came to a standstill, and the elderly woman in the passenger seat, who had dementia, kept saying: *The lights are green, why can't we go?*

It would take them roughly three hours to make it three miles. They slowly inched their way up the mountain, on the one escape route that led north from town rather than south, past the reservoir, the Calvary church, and a pizza shop. By then, everyone in the car desperately needed to use the toilet. When they couldn't find one, they stopped in the woods. Skye walked off and tried to relieve herself while keeping hold of Jude.

As they resumed their journey to Chico, Skye realized her phone had slipped out of her pocket. She had no way of reaching her father. And if he called to check on her, no one would answer.

Five miles south of the Sedwick home, on Copeland Road in Paradise, Christina Taft was woken at 8:30 a.m. by a knock on the door of the cozy, two-bedroom duplex that she shared with her mother, Victoria. She heard her mother open the door to their neighbor, a retiree named Alice Blair. Christina couldn't make out the conversation, but Victoria soon relayed the essentials: Blair's granddaughter had been driving from Paradise to work in Chico and saw the column of smoke rising from the hills behind her. Blair had come to tell them to get out.

It was hard for the Tafts to gauge just how serious the situation was: they had not received any shutoff warning calls from PG&E or evacuation alerts from the county, which Victoria, who tended to defer to authority, might have been inclined to heed.

Victoria called the phone company to settle a bill because she worried they were about to be cut off. And Christina took a shower for half an hour or so, until she smelled smoke. Outside, she realized, it had grown darker, and the traffic on the main road was bumper-to-bumper.

Relatives say Victoria once led a glamorous life on the sidelines of Hollywood and the Southern California music scene, photographed with Elton John, helping to run the fan club for The Doors, and donning a white fur or elegant dresses, with her blond hair in an impish bowl cut. Victoria's younger half brother, Donn, remembered getting a lift with her through Hollywood one day, when she pulled over to chitchat with a man watering his lawn. It happened to be Rod Stewart. She won small roles in TV and films, and was a stunt double in the 1990 movie *Dick Tracy*.

Although Paradise seemed liked the boonies to Christina, who was attending the business school at Chico State, Victoria appreciated it for its sedate beauty and safety, and for the friends she made at the Lions Club and at church. Her health presented challenges: her eyesight failed because she didn't want to use the eyedrops prescribed for her glaucoma, worrying they would turn her blue eyes brown, a possible side effect. And she sometimes had to walk with a cane, though she loved to take a daily stroll to Safeway, just down the road, especially when there was a $5 Friday sale on.

Victoria pulled a small case onto her bed and began to pack. She looked for a flashlight, expecting an outage, and directed Christina to grab this and that: pillows, blankets, tubs of documents and jewelry and coins, the squash soup that Victoria liked.

As Christina went back and forth with belongings to the car parked in front, her anxiety became shaded with anger and panic. Victoria did not seem to grasp how serious the situation was, and Christina was terrified that she couldn't get through to her.

"I'm leaving," Christina announced, thinking back to when she was little, and these words from her mother would jolt her from whatever she was doing so she wouldn't be left behind.

"So leave," said her mother.

For a moment Christina, stung, was speechless.

Victoria reminded Christina that if the power went out, the phone would only work for another forty-five minutes. But she said she wanted to wait until noon before deciding about evacuation. To Christina this made no sense: there was a gathering storm, and only a small window of time to escape it.

Standing at the front door, Christina looked back at her mother, still with the phone to her ear, in the kitchen now. Her mouth was open as if in surprise or confusion.

"Listen to the voice of reason," Christina appealed one last time.

But Victoria did not respond, and Christina left.

Many Paradisians decided to stay behind that day, because their conception of what was possible was circumscribed by their experience of all the fires that had endangered town and been defeated.

Less than a mile from the Tafts, a Nebraska-based carpenter and one-time firefighter named Andrew Duran had risen that day with only one item on his list: to clean the pine needles that had piled up around his mother's home. "If you light one match to that, it's like goodnight, Irene," he said.

Duran had spent the last few months building homes in the Bay Area, and now was visiting with his mother in Paradise for a couple weeks before heading back east. It was a routine he had: three months a year he came west for work, after which he'd fish, camp, pan for gold, and hang out on the Ridge with his siblings. This time, though, the trip weighed a lot heavier on his heart. Duran's brother Joe, a Chico fire captain, had died in September of a cancer presumed by the family and his employer to have been caused by decades of exposure to smoke and fire engine exhaust. The funeral had taken place six weeks earlier, and Duran and his mother were still mourning the loss of the man he described as the greatest person he'd ever known.

The family had moved to Paradise from the Bay Area, where they had worked as fruit pickers, when Duran was thirteen. It was a gorgeous town but Duran said that during his childhood he had seen widespread racism. Kids at school had called Duran, who was Mexican-American, *taco* and *burrito*. He developed what he called an attitude problem and got into trouble as a teenager, first with drugs and then for an armed robbery. During a stint in juvenile detention, he had the opportunity to go to inmate fire camp and learned the fundamentals of firefighting. In a way that had made the whole sorry experience worthwhile.

Waking at 8:00 a.m., Duran pulled on his brown Carhartt overalls, splashed water on his face, and was headed toward the door when his seventy-nine-year-old mother came inside from her perfectly manicured garden. Her home was a small converted garage that the family had owned since 1980. The Durans loved living on Deerpark Lane, a tight-knit street that saw almost no traffic. "Good morning, ma," he said in the cowboy twang he'd picked up in the army that had never quite left him.

"They're evacuating the city," she said. "It's mandatory. The city is on fire."

Thick smoke was wafting through the air as though a giant bonfire was burning nearby. He hurried inside to pack his things, most importantly the brand-new green DeWalt tool bag containing the hammer and saw that he needed for work, when he paused.

Where are you going? he thought to himself. His mother would have nowhere to live. He knew how to fight this. Joe would have been mad at him for putting himself at risk, but this was what he himself would have done. *You got to stay here and save Mom.*

She begged him to go with her, and told him he could die if he didn't. *I'd rather die than see you homeless*, he thought. Duran was stubborn and unflappable. "You either got to shoot a gun or stab me for me to get concerned," he'd say. His mother knew there was no way of getting him to change course now that he'd made up his mind. "God bless you," she said, putting her hand to his cheek as she said goodbye. Reluctantly she left.

He heard the fire before he saw it, the *whoosh whoosh* of it moving toward the neighborhood. Angry winds cleared some of the thick smoke, and overhead the sky was a martian red.

Neighbors yelled at each other to leave as they ran to their cars

and sped off, tires peeling out. Two other men were preparing
to stay, but when their house lit up, they fled, too. As they drove
away, Duran gave them a thumbs up.

Before long Duran was alone on Deerpark Lane, and he started
watering the grass and trees around his mother's house. Armed
with the hose, he had just climbed up a ladder and hopped onto
the roof when a voice in his head told him to turn around. A wall
of flames 30 feet high was bearing down on him from the next
street over. Duran tried to wet the roof, but nothing came out.

What the heck, he thought, his dark brows furrowed as he stared
at the hose. *Oh, you're in big trouble now.*

As the water dried up—probably, he thought, because so much
of it was being used elsewhere in town—so did his moxie. Arriving on the street of two dozen homes, the mass of fire was as loud
as a "fleet of trains." It ensnared house after house, the one across
the street, his brother's van. There was a *pop pop pop* from the
home of a gun enthusiast: bullets igniting in the heat of the fire.
"I get to thinking right there like maybe this wasn't a good idea
after all," he said. His mother's combustible little cottage was an
ill-advised vantage point. He offered a quick prayer to God: *You
gave me the foresight to stay and I know You're not gonna kill me now.*

Duran took only what he could carry: his duffel bag. He hid
his tool bag under an overturned wheelbarrow, hoping it might
survive the flames. He wanted to leave, but because the fire was
burning on both sides of the street, hopping from structure to
structure, he decided to take refuge on the fire-resistant asphalt
road. He would wait until it burned past him.

Burning pine needles sprinkled down on the neighborhood
like ticker tape, starting a fire on the fence along the property
line. The flames crawled along the barrier, and then ignited the
giant pine behind his mother's home, and in minutes the house
itself. Embers found openings, and it started to burn from the

inside out. *It's time to get out of here*, he thought. In Duran's mind, you'd have to be a guy with 10 arms to stop a fire like that.

His other brother's home was just a few houses down, and he had to watch that burn, as well. *None of these people would have ever thought in their wildest dreams that it would end like this*, he thought. He remembered all the good times he'd had in this neighborhood: his siblings' birthday parties and the weekend barbecues he hosted on his visits that he called "chicken Sundays." All gone.

"This is the new Deer Creek Park," Duran shouted in a video he made on his phone showing a neighborhood engulfed. In the chaos of the situation, he mixed up the name of the street his family had lived on for decades. "There's my mom's house, she's going up right now—sorry, Mom. There's your neighbor, sorry. Mike's house—gone. I'm right here letting it blow on by so I can get a clean path to walk out through. So far so good. Whole street is on fire. Love you . . . bye."

After the fire had burned over one side of the street, Duran picked his way through the blackened landscape. He headed right from Deerpark Lane, past fallen electrical wires, and then to Middle Libby. The road was empty and dark, but eventually through the smoke he spotted a pair of headlights in the distance. It was a fire truck from nearby Chico—in fact from the very same department his brother had worked at for nearly thirty years.

What are you doing out here? an astonished firefighter asked.

I was tryin' to save my mom's house, he said.

Any luck?

The firefighters were all over: in central Paradise attempting to rescue hundreds of residents who had been stuck in traffic at Wagstaff and Skyway and had to flee their cars to escape the flames, in Butte Creek Canyon getting eyes on the fire that was now headed toward the town, and at Feather River hospital.

"Priorities are evacuations and public safety first," incident-command officials radioed to crews.

At the far north of Skyway, just before the road left Paradise and entered Magalia, Cal Fire chief Ken Lowe had never seen a fire so intense. It was almost as if it had an intelligence, a will, of its own, moving just beyond his clutches. As residents like Christina Taft and Andrew Duran tried to flee, Lowe had been trying to get in with his "strike team" of five engines, which had three firefighters each.

Near the Fastrip gas station, as vehicles began to catch fire and their occupants fled with nowhere to go, Lowe and his crews did the only thing they could think of to save lives. They shouted for people to run toward the wide intersection.

There, Lowe's five fire trucks formed a physical barrier around a hundred or so civilians and their dozens of dogs and cats, protecting them as the flames went past. "It was panic," Lowe said: pitch-black, trees lighting up like fireworks all around, power lines arcing, embers skidding through the sky like cherry blossoms on a gusty day.

In southern Paradise, on Carroll Lane, Bill Goggia awoke to a poisonous orange atmosphere so thick with smoke he couldn't see the sun. He heard the piercing metallic clang of propane tanks exploding in the distance. His sister, who lived nearby, called to ask him to help a relative in the area, but Goggia told her that he couldn't: chunks of burning wood were falling from the sky.

Goggia, a locksmith for thirty years, lived in the same stucco, three-bedroom house he grew up in. Now it was time to leave. By 9:00 a.m. he was on the road, accompanied by his tabby cat, Mikey. But the street was so clogged with people trying to escape that Goggia barely moved. The fire was getting closer. The van was going to burn up with him and the cat inside, he thought. So he turned back against the traffic and headed toward the home of his neighbors. On a rocky patch they had erected three enormous crucifixes. That was where he decided to leave his van with the cat still inside. He addressed God: *I'm gonna leave him in Your hands.*

In the suffocating smoke, Goggia went back out and followed the road by foot, hoping it would be easier to make progress. But at fifty-six, he was not in perfect health, and he had suffered a heart attack three months previously. He hadn't advanced very far before he was overcome by the choking fumes and collapsed to the ground. Vainly he tried to get up or get the attention of passing vehicles. A woman yelled at him to *get the fuck out of the road.* An ember burned his left eye so badly he couldn't see out of it. Lying in a ditch at the side of the road, he thought he was going to die.

Finally, a truck stopped for him. "Hell yeah, you can get a ride," said the two men who crammed him among their belongings. Instead of aiming for safety, however, they turned toward the flames: one of the men wanted to save his tiny home, a residence the size of a garden shed that could be transported on a trailer.

Goggia thought the man was out of his mind, and, indeed, just as they reached the tiny home, it caught fire, and the man retreated.

It seemed like a fatal detour. The road out was now blocked by flames, and the driver seemed unsure what to do. His friend pulled him out of the front seat and took his place, but the car would not start. Flames coiled and flickered into the cab through the open windows, and the sides the truck began to melt. The floorboard was so hot that they had to lift their feet. Goggia's leg nearest the door was severely burned, and he poured a bottle of water over himself to stop himself being cooked. Fires were igniting in the cab and he tried to stomp them out.

Finally, the engine kicked in. The driver jammed the accelerator, and they blasted through the wall of fire.

In a town with so many elderly and disabled, the challenge for many was not simply getting out of Paradise. It was getting out of their own homes. Not far from some of the refuge areas that firefighters and police created on the fly, Andrew Downer was marooned in his living room, surrounded by the uranium glass and marbles and all the other antiques he and Iris Natividad had so lovingly collected over the years.

Natividad had left a few days before the fire for her office three hours away, in Santa Rosa. Downer had gotten a call from a friend at 7:37 a.m. to tell him about the blaze, and to offer him a lift off the Ridge. Downer said that he'd just woken up and the fire was far away. In any event, he was only going to leave if the situation was desperate. The friend said that he'd drop his family off in the valley and come back for Downer. Natividad called Downer from the road, and agreed with his decision to wait it out. At that early hour, with the fire still east of town, it didn't sound like it posed a threat.

Just after 8:00 a.m., Natividad got a call from a friend who lived by the canyon on the east side of Paradise. Firefighters had knocked on his door and told him the fire was coming up the canyon and that he had five minutes to leave. He was in tears— he was going to lose everything, he told Natividad; he wouldn't have a house. Feeling bad for him, Natividad tried to lift his spirits. The friend told her that he and his family were going to his parents' place on the other side of Paradise, and offered to collect Downer on the way. Iris told him that he didn't have to worry because somebody else had already volunteered. Still, calling Downer again, she wondered out loud whether she should turn around and come for him herself, even if she couldn't quite believe the fire would cut through swathes of neighborhoods and reach their home in Paradise's little downtown.

By this point, Downer was almost completely reliant on others for something as simple as leaving his own home. It had been a terrible week for Downer, when the true extent of his disability had become apparent. Antiquing was impossible. Forget getting in and out of a car—it was difficult merely to make it to the bathroom by himself. He was utterly humiliated. *I'm a 54-year-old man*, he told Iris. *You have to bathe me. I'm like a baby. This is not what I want. I'm a burden to everybody.*

It was in Natividad's nature to help, and she loved him, but the strain of full-time caregiving whenever she was at home in Paradise was wearing on her, as was Downer's grinding unhappiness. Yet it had seemed to both of them that, just recently, things were improving. A caregiver reminded him that if he stopped drinking he could figure out how to walk on his prosthesis. Downer's new therapist was working on his anger and trauma. The physical intimacy between Iris and Downer was gone, but the love remained. "It's not that romance love, it's that caring for somebody love. You're with somebody for 28 years."

There were a number of calls between 9:00 and 11:00 a.m., when it became apparent that the options available to Downer had narrowed. The friend who was evacuating to his parents' home called to say that his parents were evacuating too. Downer's caregiver, who lived only a couple blocks away from Downer, found that the traffic was backed up and only flowing in one direction, away from Downer. Calling from Chico, the man who made the 7:37 a.m. call said he was facing the same problem: he couldn't drive back up to get Downer because there was gridlock.

You have a car there, one caller told Downer. *You can't tell me you can't get into that car. That car is right there. Skyway is right there. You can just get in that car and drive away.* But Downer replied that he simply could not. He was physically unable to get out of his wheelchair.

Natividad was in what she calls her "crisis mode"—she had learned, through her stressful car-dealership job, how to tuck her emotions away and focus on the practical. Call 911 before the lines go out, she kept telling Downer. The firefighters will come and get you. He relayed what they had told him: four hundred people were trapped, and it didn't look like they were going to make it to him. She told him to go outside and get ready for the fire department. Make sure the doors to the backyard are open, she said. If the firefighters can't take the two dogs, they'll have to fend for themselves. Have your pills. When you get down the hill, call me. People are waiting to take you in.

Downer's anxiety seemed at a low ebb, which was unusual. He was terrified of falling over, for instance, his height making it that much more calamitous, and often when paramedics came over to help get him to a doctor's appointment, and they were lifting him out of his chair, he would panic and start calling for her: Iris, Iris, Iris. But today he, like Natividad, was keeping it together. He never told Natividad what he could see out the window, but toward 11:00 a.m. he said he could hear explosions, and that they were getting closer.

"Maybe this is a good day to die," he suddenly told Natividad. He meant, Iris gleaned, because of his brother Michael Downer, who had been killed about thirty years before, almost to the day, when he was driving in fog, missed a stop sign, and went underneath a truck at an intersection. Natividad batted the comment aside. She did not believe for a second that Downer would perish. A problem had presented itself and she would solve it. That was what she did.

Soon, as she expected, they lost contact when the lines went out.

As escape became impossible, firefighters across the Ridge adopted the makeshift tactic, called "sheltering in place," that Cal

Fire chief Lowe had used near the gas station. At one location, people were corralled into an intersection as firefighters aimed cannons at the encroaching flames, like soldiers defending a position from attack. But there was little water to spare, because thousands of yard risers and domestic water lines had melted, resulting in widespread leaks, and hydrants had run dry.

For two hours, Paradise police officer Rob Nichols had been directing traffic at the same location as Lowe, the corner of Clark and Skyway, and had already helped evacuate a pregnant woman who had gone into early labor. The woman, Anastasia Skinner, suffered from a medical condition that made her situation precarious: if she gave birth outside of a hospital she would likely bleed to death. She was able to leave with help from the first responders. But by 11:30 downed power lines and abandoned cars made the roads impassable. This was not a spot to get trapped. The Reliance propane yard was just a block away, and across from where Nichols stood, in the parking lot of the old Optimo Lounge, was the Fastrip gas station, with its own massive propane tank. To his right was Fins, Fur and Feather Sports, a sporting goods store filled with live ammunition. "The last thing we need is bullets flying," he told firefighters.

The fire was getting closer. One woman was so desperate to escape it that she raced into the Optimo parking lot and ran over a row of traffic cones. Another stood on the curb sobbing as she watched a home across the street burn. It was hers. At a house down the street the fire sent flames spewing out of a propane tank, and panicked evacuees fled toward the first responders.

The town had two designated assembly points where people could seek refuge if they became trapped. The Paradise Alliance Church and the senior center were large buildings with spacious parking lots for residents to wait out a fire in a sea of nonflammable asphalt, but those were miles away. Over the radio, Nichols could hear that traffic was backed up all over town, and Skyway was being overrun with flames. Blocks ahead, he saw helicopters carrying buckets hovering what appeared to be just above the ground. *There's no point in sending people down the hill if they can't get out,* he thought to himself. With the roads to assembly points, and those out of town, blocked by flames and fallen tree limbs and power lines, Nichols and others decided to create an impromptu refuge.

The best option was across from the gas station—a parking lot with a cluster of tin-roofed stucco buildings near the Optimo, popular in town for its Chinese food and karaoke. Crews used hatchets to gain access to them. One was supposed to be a coffee shop. Nichols used his carbide spring tool, the same device he'd wield to shatter windows and rescue people in car accidents, to break a glass door, and more than a hundred people were hustled inside with their animals. Some were on crutches and in wheelchairs. The police, who had spent the day so far directing traffic and trying to get people out, had to slow down and tend to the vulnerable residents.

Nichols and his colleagues stood with worried evacuees, making conversation or helping them plan for when they could eventually leave. They tried to keep people comfortable as the hours passed. One volunteer officer found the door of the Optimo was unlocked and guided residents inside to use the bathroom. Firefighters sprayed down the bar to protect it from embers landing around it. Nichols asked around for cookies or candy for an elderly diabetic woman with low blood sugar. The Optimo had plenty of alcohol but limited food, and it didn't seem right for a cop to break into the convenience store across the street.

Nichols had already come up with a back-up plan in the event that they had to abandon their refuge near the Optimo. Earlier he'd hopped into his police cruiser and drove up to a storage facility—a maze of asphalt roads and small metal-roofed buildings—immediately behind the Optimo. His car had a large bumper on the front designed to move vehicles out of traffic, and he drove slowly into the gate, gently tapping the gas until it broke open. If the fire overwhelmed their safe zone, this was where they'd come.

Among those hunkered down near the Optimo was Erin Roach, a fifty-eight-year-old retired waitress and amateur artist. Everyone, even the dogs, seemed to her to be remarkably composed and self-possessed despite the fire outside. In the panic that morning, the only things she'd thought to retrieve from her home were her purse, the nebulizer and medication she used for her asthma and COPD, and her acrylic paint set.

Earlier, driving toward Paradise from the north, Roach had gotten stuck in traffic at an intersection and stepped out of her car. First responders had commandeered a Pepsi truck that was also gridlocked, separated the tractor unit from the trailer, and employed a chain of volunteers to empty the crates of soda. This was to be a sanctuary for Roach and others in case the fire blew over them. Instead the jam cleared, and she got back in her car and proceeded as far as the Optimo, about 1,500 feet down Skyway, when she was ordered out and into the parking lot. Firefighters were hoping for water drops, and told the crowd to get on the ground when the aircraft flew by in case they were knocked over by thousands of gallons falling from above. But it was too smoky for the helicopters. Soon Roach did what she'd always done since her daughter became seriously ill with an autoimmune disease and Roach needed "to keep my mind from thinking too much." She started to paint: a sky filled with rough blotches of color that suggested ash to her. "It looked like rain coming down of red and blue."

They were in the safest spot they could be, Nichols thought, but people were worried. In front of the Optimo a woman had asked Nichols: "Am I going to die today?"

"You're not going to die today," he told her. "You're going to be fine."

Within an hour of Skye's departure, John Sedwick had left his house and began helping neighbors prepare for the approaching inferno. One of them was Waylon Shipman, known as Bubba, a rugged, cowboy-hatted forty-three-year-old electrical contractor who lived in a green-roofed two-bedroom next to the graveyard. Shipman had parked his truck, trailer, Harley, 1967 El Camino, and 1963 Chevy Impala in between patches of green grass at the cemetery, the only place he thought they would be safe.

Near the smoke-cloaked headstones, under the swaying pines, Sedwick, with a shovel hooked on his back, stepped out of the haze around 10:00 a.m. and offered Shipman a hand. He told Shipman that he used to be a firefighter, and that he lived just behind the Depot restaurant. Shipman knew the house—he admired Sedwick's impressive midcentury Isaacson bulldozer every time he drove by. Sedwick had bought it thirty years before with an $1,800 loan from his niece and her husband. He'd used it to keep his land clear, and to build a proper road from his house to the Depot after his neighbor had blocked off the original driveway.

Sedwick and Shipman chatted for about ten minutes. Shipman's firefighting experience came courtesy of a two-year prison stint for manufacturing meth. He had started using the drug in his twenties while building houses; it gave him the energy that the back-breaking labor required. After he and a coworker lost their supply, they started making their own. It eventually landed him in prison, where he spent his sentence as an inmate firefighter battling more than forty fires across the state.

When Sedwick said goodbye, Shipman noticed his unusual eyes: hazel rimmed with white.

Later, as exhaustion set in from hours of chopping down vegetation, Shipman began to wonder if this encounter had really happened the way he thought. Why would an elderly man have appeared in this deserted town and come to his aid? Shipman was a churchgoing man, familiar with the myths of religious visitations, annunciations, and revelations, and the thought occurred to him that maybe he had met an angel.

6

The Cemetery

By the afternoon, entire portions of Skyway were little more than burning rubble. Throughout Paradise, smoking foundations were strewn with unrecognizable piles of warped metal. Staircases leading nowhere and chimney stacks denoted homes. Neighborhoods along eastern Paradise were completely wiped out, creating sweeping new vistas from the roadway. Hundreds of charred cars lined the major arteries in town.

To the north, fire had destroyed entire streets in southern Magalia, and Ridge View high school had burned to the ground. Authorities rode through neighborhoods with a loudspeaker ordering residents to flee. But the Depot still stood.

Nearby a retired Cal Fire captain named Lloyd Romine was trying to calm a panicked resident. Romine was pretty worked up himself. This was the fire he'd always feared would rampage through the Ridge. It was consuming homes just a few dozen feet from where he stood. A handful of people with shovels and rakes stood in front of the houses that remained, trying to bury the falling embers before they set fire to their yards.

At around 2:00 p.m., an elderly man in a silver Isuzu van pulled up to the stop sign at the intersection next to the cemetery. A look of recognition flashed across the driver's face and Romine walked over to him.

"I'm John Sedwick. I used to be a firefighter for you— Company 34," he said, his eyes shaded by a silver helmet. In the hours since he'd met Bubba Shipman, Sedwick had driven the winding Magalia roads near his home—Indian Drive, Ishi Drive, and Yahi Court—to see which areas had been impacted by fire.

Romine smiled: "John, I know who the hell you are."

A firefighter in Magalia for about a decade of his thirty-year career, Romine's resume also included stints teaching, conducting arson investigations, and working with inmate crews. He sometimes jokingly described himself as a "hippie with a badge," more liberal than most in the fire service and concerned about climate change. He and Sedwick had worked together for five years in Company 34, where Romine was Sedwick's supervisor. Sedwick, a volunteer, always showed up, even when others were too busy, and he had a knack for beating Romine to the scenes of collisions and fires. He recognized the helmet Sedwick was wearing. Two decades earlier, Romine told him it was no longer up to code—he needed to wear a plastic helmet, which provides better protection from live wires. Sedwick had retired a few years into Romine's tenure.

Despite the years that had passed since then, Romine could see that Sedwick was in his element. "It's looping around the canyon . . . Ishi Trail is gone and all that area is gone," he said, serious but smiling. He made Romine think of a race horse just let out of the block. "This is the big one," Sedwick said.

"We've gotta figure out something," Romine said, bending his 6'2" frame down to the window.

They agreed Sedwick would work recon for a group of nearby firefighters, because he knew this area like no other. It would be safer than having out-of-towners try to maneuver large engines down narrow isolated roads they weren't familiar with—it could even save lives. Romine was worried about the unpredictability of the fire. He thought they had no reference point for its size, and the way it seemed to reverse direction every few hours.

As they were talking, Sedwick suddenly reached up from the driver's seat and tenderly patted Romine on the head.

"Your hair is starting to catch on fire," he said.

Anna Dise and her father Gordy lived on two and a half hilly acres in Butte Creek Canyon, a settlement tucked in one of the valleys that girded the Paradise Ridge. Their property was dotted with old cars Gordy worked on—a Jaguar, a Corvette—as well as his wire-fenced enclosure of medical marijuana plants and a vegetable garden of zucchinis and tomatoes. It was gloriously situated, with vistas of stark canyon walls and a rock outcropping that suggested a man's face in profile. Amid this majesty, the Dises faced difficulties—there wasn't much money to go around, for instance. And the home lacked electricity for long periods in Anna's younger years.

Yet Anna was not one to complain about it, and developed both strength of character and a remarkable hardiness. On the coldest of days she was happy enough in a T-shirt, her eyes a pale blue and her cheeks pinked. She was inspired by characters and ideas from the novels she devoured and, occasionally, wrote. Her beliefs were her own. When her father posted on Facebook criticizing the Supreme Court's legalization of gay marriage in 2015, which he said was because "there was too many females on the court," Anna replied: "Everyone has opinions and is entitled to them, just don't try to force your opinions and views on other people. On a happier note I finally get to see some of my friends get married."

News of the fire had reached them early, and they had begun watering down their property to reduce its flammability and using a weed eater to chomp through a neighbor's undergrowth. For hours, though, the drama unfolded in far slower motion than elsewhere.

At 1:02 p.m., an engine reported seeing fire on both sides of Butte Creek Canyon. At 2:00 p.m., night seemed to fall, and they couldn't find their way without a flashlight. Gordy wanted Anna to cook dinner—chicken and broccoli—at 6:30 p.m., and because the power was out she did it in the dark. After they'd eaten she went outside to look for the blaze, and saw that it was already plunging from Skyway, which traced the eastern canyon wall, toward them.

Anna and her father began grabbing belongings from their trailer: paperwork, books, computers, Gordy's motorcycle shirts. Gordy had wanted one last thing, and he ran back even though the deck he had built in front was catching.

From the pathway, Anna yelled for him and tried a car's horn, but he did not reemerge. As she watched, one wall collapsed toward her in a flaming mess. The vehicle next to her was catching fire and so, still honking, she tried to back the car out of the driveway. But the wheels had started to melt. She and the dogs jumped out and she pulled out what she could, though there was only time to get her purse, a photo album, and her dad's laptop. It was too dark to see what she was reaching for, and as soon as embers got inside the car, the whole thing went up.

Anna hurried down the driveway to the rutted road that passed their property. By the side of the road was a ditch a few feet deep, and Anna wetted her clothes and her dogs. Incredibly, despite the lack of rain, water was flowing through it.

When she called 911, she was told someone was coming for her, but no one did. Every time the wind gusted it tossed embers into the air, so Anna continued to splash water over her clothes and her hair. At some point in the hours that followed her dog Sirius vanished.

Paradisians were completely unable to reach relatives and friends living only a few miles away.

Chellie Saleen had risen at 6:00 a.m. and observed a pall of smoke. She received an alert by phone, and "all of a sudden someone was banging on door—it was my neighbor saying, 'We gotta go now.'" Saleen changed out of her nightgown into jeans and a shirt and tried to call her seventy-five-year-old mother, Joanne Caddy, who lived in Magalia, but couldn't get through. Her mom turned her phone off at night, and took Ambien and Trazodone to help her sleep. The previous evening, as she had left the house filled with antiques, wooden furniture, and china, she had told Caddy she loved her. Caddy was tucked up in her favorite spot, her bedroom, a comfy nook that was easy to keep warm, watching television.

Her mother didn't drive and depended on her for everything from her shopping to her medication. She was, Saleen said with a smile, her mother's "honeydew"—"as in, honey do this, honey do that." Gentle and with a ready, throaty chuckle, Saleen was always in the company of Poppy, a small bird the color of popcorn, who sat on her shoulder, blew kisses, and said "Hi, baby." Caddy had dementia, but she could be a hoot, dancing to music in the supermarket and telling schoolgirlish stories from her youth.

Saleen threw a few things in the car and headed in her mother's direction through the abysmal traffic, but she was blocked by "a wall of fire." She called her son and told him to keep trying her mother. In the stop-start procession of vehicles, Saleen's car overheated, so she restarted it to no avail, then pulled over, popped the hood, and called out to the cars creeping past: *Does anybody have any water?* Two guys helped her, but it overheated again, and by the side of the road she tried to call her mom once more. "It was stifling hot, it was hard to believe it was 9 in the morning," Saleen said. "People were walking down the street with dogs, people were on bikes. Emergency vehicles couldn't get through. The whole time I was thinking about her—but I also had to be present because it was so dangerous." Poppy was perched quietly in his cage; she worried about him being unable to breathe. Finally she got off the Ridge.

Meanwhile, Pat LeBlanc and her husband, Tom, who lived in southeastern Paradise, just off Pentz Road, had been packed and ready to leave on a trip to California's North Coast that morning when they were told first by a neighbor and then by a robocall to evacuate. They canceled their house sitter, and when they went outside they saw ash coating their cars and heard the distant roar. Kimber Wehr, Pat's daughter and Tom's stepdaughter, lived near the center of Paradise, only a ten-minute drive away, and when Pat called her she sounded like she had just woken up. Pat told her daughter to be ready to leave and to keep trying 911, and it wasn't until Pat and Tom got on the street in separate cars that they realized they would not be able to make it to Wehr, who could not drive.

Wehr's biggest fear in life was becoming homeless, so the LeBlancs had bought a home in Paradise that Wehr rented from them for a nominal sum. She was not the kind of daughter that Pat had expected; Wehr didn't seem to want to wake up or go to sleep at the same time as most people. She didn't enjoy family gatherings, and disliked looking in the mirror or having her photograph taken. Still, she adored her mother's impromptu invitations for pizza, as well as the Paradise library. They saved the newspapers for her, and she devoured audio books. Although she didn't have much herself, at Christmas she left kids' books in waiting rooms and churches around town.

LeBlanc and her husband drove out toward Chico, not dreaming the fire would spread particularly far. But after a while the phone lines went down and they couldn't speak with Wehr anymore.

Other trapped Paradisians had relatives who lived or worked much farther afield, and as the afternoon turned to evening the hours became laced with anxiety as they waited to hear what was happening inside town.

Lonnie Walker, a truck driver who hauled logs, was stuck outside the Ridge with no way of knowing what was happening to his wife, Ellen. That morning, he hadn't wanted to wake her: she couldn't get around easily owing to fibromyalgia and migraines, and was sometimes confined to bed for days at a time, though she always made sure to greet Lonnie at the door when he returned from work. She'd just had her hair done, and the whole process of going into town could exhaust her. She still had her hair scarf on. It was blowing a gale outside, but Lonnie thought it would be just another windy day.

Lonnie had met Ellen thirty years before, after they both had other marriages, through a dating service advertised in the newspaper. "We made arrangements to have lunch together and to be honest with you I just admired her dignity, her ladylike manners, the way she carried herself," Lonnie said. "Quiet, she was extremely quiet, extremely courteous, and she taught this old man a lot of good attributes that he should have had a long time ago."

They were tickled with the home they found in Concow, a two-bedroom with a lovely back porch and a two-car garage. Lonnie was always hankering to move to woodsy Oregon, but this was a perfect substitute. Ellen adopted cats, planted roses outside, and kept her weekly appointments at the hairdresser, when Lonnie would wait for her and they'd go for a meal afterward. She was proud of her short, blond curls. "I remember I'd come home from work and her hair wasn't just what she thought it should be," Lonnie said. "She'd put both hands up there and look at me and say, *I know I don't look very nice to look at.* And I said more times than one, *Honey I didn't marry your hair. I married you."*

Raised amid cotton fields and grape vines in the Central Valley, Lonnie had been a trucker for the best part of fifty years. He didn't have to continue working at his age, but he wanted to, because it brought them a bit of spending money, for eating out or garden plants. Usually when he left in the middle of the night, she would wake up, too, and as he was walking out the door she would tell him to *never give up.* That was one of her favorite songs: *Never give up, Jesus is coming.* On November 8, he only had two days left until he was due to retire and could stay home with her.

Lonnie was 100 miles from home, in mountainous Plumas county, when he heard about the fire, and when he drove back there were roadblocks set up around the Ridge and he couldn't get back in. He called the sheriff's office and told them that Ellen was disabled, and that they needed to get in there and help her. He hoped she was able to look after herself amid the dislocation.

Over 500 miles away in Portland, Oregon, Angela Loo was trying to call her father, Ernie Foss, and his stepson, Andrew Burt, who lived together in Paradise. For Foss, Paradise was the last stop on his protracted exile from San Francisco, the city in which he was born in 1955. He was molded by the Summer of Love and the hippie movement, gigged all over town with dreams of rock stardom, and, according to his first wife, met Jerry Garcia and Janis Joplin. When he grew tired of eating beans out of cans, as he put it, he got a job as a studio musician. By twenty-three, he had two children, and was also taking care of his younger brother because their parents had died.

"He gave up that lifestyle for us—he said he wondered where he could have been if he hadn't," Loo said. "Not that he didn't drag us from gig to gig," like the long-running Exotic Erotic

Ball, to which Loo was first taken as a wide-eyed preteen. By the 1990s, Foss, a mellow, heavyset man, always sporting a reddish mustache or beard, was working at a New Age jewelry store called Touch Stone, and he also gained some renown as a tarot card reader. A couple years later, Loo noticed his health problems. Foss had loved to surf and skateboard, but on a trip to the theater with her dad when he was about forty, he stopped every few doorways to rest because of pain in his legs.

Eventually he was diagnosed with lymphedema, a disease in which fluid builds up underneath the skin and causes swelling. In its most severe form it leads to elephantiasis, and at his worst, Foss's weight soared to over 500 pounds, and each of his thighs measured over 50 inches in diameter. His stretched skin caused such agony that he couldn't sleep, and he had to take a cocktail of pain medications. His third wife, Linda Foss, a former animal keeper, was also disabled, and they couldn't make rent.

Together with Linda's adult son, Burt, they found a place to live in Paradise. Although it was far from the Bay Area, the air was clean, there was a hospital close by, and the two-bedroom house had been built by a man especially for his disabled parents. By now, Foss was bedbound, and as Linda's renal disease and leukemia progressed, they spent the days next to each other on hospital beds in the large, tiled living room. When Linda was taken to the hospital for the final time, Foss couldn't get out of the house to be with her, and had to say goodbye over the phone. At least he still had Burt, a baby-faced man in his thirties, who bonded so much with Foss that he started to call him "dad."

Burt looked after his new father, which the state of California also paid him for. He cleaned Foss's body to make sure he didn't get skin infections, massaged his feet, and took care of his bedpan. The pair watched James Bond and Doctor Who, played online games with family and friends, and shared an appreciation for music, documentaries, guns, and weed. They had a beautiful dog called Bernice. You couldn't call one of them without the other hollering in the background. "It really was like the odd couple," said Loo. "Just two bachelors living on their own with no one to watch them."

Foss had volunteered with a local citizens' wildfire safety organization, monitoring radio scanners from his bed and participating in an online community. He knew Paradise was at risk of wildfire, and worried about it "going up like a matchstick," Loo recalled. But the town was good for him, not least the hospital. "Honestly, even though his health wasn't great, it was the best care he ever received," Loo said. The staff who provided his routine treatment at Adventist Health Feather River were attentive and kind, offered in-home services, and helped Foss maintain a measure of independence. In Paradise, Loo said, "they felt like they had finally found a refuge."

They didn't have cellphones because service was poor, and at 8:00 a.m., when Loo first tried their landline, it rang and rang. At around 9:00 a.m. there was a dreadful, squalling electronic sound, like a fax machine. Websites showed that the zone in which the pair lived had been evacuated. She rang everyone she could think of for help or information, and got others to do the same.

Lighting a candle next to a photograph of her brother, who had died of a sudden heart problem a few years previously, Loo offered a prayer to him: *If you have any strings to pull. . . .* Waiting at home for news, she walked over to the stove and thought she heard her father saying "Angela" just over her shoulder. Whipping around, she saw only an empty room.

At the same time as Chellie Saleen, the LeBlancs, Lonnie Walker, and Angela Loo were hoping for news of their loved ones, another Paradise resident named Greg Woodcox was making a three-and-a-half-minute video in town, a half mile from the home of Foss and Burt. Woodcox had been caravanning with several other cars through a residential neighborhood on the verge of being engulfed when their path was blocked by a fence. He jumped out of his truck, leaving his two dogs behind, to suss out a way through when the flames exploded all around him. A fox ran by, and he followed it. It led him to a creek, where Woodcox crouched in shallow water until the worst had passed. When he reemerged, he started filming. What had been a normal street had become a monochrome moonscape. And the people in the other cars were skeletonized, or were unidentifiable mounds of molten black. Some of these were people Woodcox knew. Remarkably, Woodcox's two dogs were fine, and his car engine was even still running. In shock, his voice shaking, he credited the Holy Spirit and drove away.

After running into Romine, Sedwick had driven all around Magalia, attempting to map out the perimeter of the fire—its head, flanks, and heel. At that point it was still roaring through Paradise, and covered more than 50,000 acres. Back at the Depot, Sedwick told Romine that the homes of a few former volunteer firefighters were gone.

The sun was a small, orange-red orb, what Romine and other firefighters used to call "the dragon's eye." Romine had seen it many times—it suggested a raging fire.

At 5:30 p.m., in front of the Depot, Romine and Sedwick said their goodbyes. Sedwick planned to return to the cemetery to monitor the fire, while Romine wanted to check on his home. He left Sedwick with a plea: "John, if things go to shit you gotta get the hell out of here, you understand?"

Nearing the end of the first day of the fire, the Ridge and its environs had been almost completely depopulated.

That afternoon, the incident command center had radioed

Rob Nichols, the police officer who had helped create a temporary shelter near the Optimo Lounge, and asked him to drive the main roads and find a path out for the evacuees. It was the first time he saw the scale of destruction. Fallen power lines crisscrossed Skyway, and it took a moment for him to stop seeking out the usual landmarks—they were gone. He found that Clark Road was, barely, a passable exit route.

Back at the Optimo, he directed evacuees toward the Kmart, and from there down the hill. After five hours of waiting, they fled from the stucco buildings to their cars in the gathering shadows of early nightfall and hurried off the Ridge.

Erin Roach, the painter, found that her abandoned car had somehow survived, and she offered a lift to a lady she'd met. By that point, Roach said, nothing could have surprised her. "This is how anesthetized you get with fire: on the way out through town, I looked over and the KFC was on fire. And all I said was, *Oh wow, that's gonna be a really greasy fire.* You just see it in flames like it's nothing and just keep on driving."

After they had all left, Nichols went to meet his colleagues at Kmart, and then he stopped at his house. He expected it would be gone, and when he got to his street, sure enough everything was flattened and gray. His own home had just fallen in and was still burning. The warped body of his motorcycle stood upright in the garage, bright orange flames just a few feet away. The jeep he took on his weekend off-roading adventures was also a shell, with blackened pools of rubber underneath the hubs.

At Clark and Skyway, embers continued to fall, and that night the Optimo burned.

John Sedwick toiled to protect landmark buildings and neighbors' homes. He managed to save the firehouse in which he had once worked. It dated back to the late 1880s, and also served as a general goods store, billiards parlor, and saloon. It had been part of Sedwick's life since he was a child, and he could describe its interior in 1945 from memory. "The first thing you would notice was the jukebox, which was to the right of the door," he wrote in a historical recollection. "It was a tall cathedral-shaped rig with multicolored bubbles floating in glass tubes." His mother worked as a cook there when it was nicknamed "Bucket of Blood," in a bid to suggest an English pub and appeal to servicemen returning from Europe after World War II. In 1957 it was taken apart and reassembled at its new location on Skyway to serve as a firehouse for the Magalia Volunteer Fire Department.

That evening he called both his daughter and his niece, Simona MacAngus. He sounded calm, and, as became clear to them, he had reason to be proud. While most residents of the area had left, he had fought the fire himself, and told his niece that he had managed to get his ancient dozer working and had cut fire line with it. He had spent the day helping his neighbors, men half his age. When he got hold of Skye around 10:30 p.m., she was in Chico, where she had finally met up with her boyfriend. It had taken her seven hours to get down to the valley. A friend had offered them a room for the night belonging to his thirteen-year-old daughter. Amid stuffed animals and coloring books, Skye listened as her father recounted: *I'm OK. Fire came right up to the door, but I fought it off.* "He felt purposeful," Skye said. "He felt like he was in his niche."

In Magalia about an hour later, Sedwick couldn't help but climb up onto the bright red 3,000-gallon water tender bearing the name of his old unit: Upper Ridge. He had volunteered at that station, putting out attic fires and responding to wrecks along Magalia's windy roads. When he saw the machine at the end of his driveway just next to the Depot Restaurant, he had to get a closer look.

"Whoa, you're in a dangerous spot," a firefighter named Ray Johnson yelled. Sedwick's presence had startled Johnson, who was gathering the hose from the vehicle. The Ridge was under mandatory evacuation, all but abandoned except for first responders, and yet there was an elderly man hanging off the side of his tender.

"I'm John Sedwick," he yelled. "I used to be a volunteer firefighter."

Johnson had fought the fire all day, and his face was gray with soot and ash. Monday through Friday, the forty-two-year-old father of two was a heavy-equipment driver for the state department of transportation, but in his free time he served as a volunteer fire captain. He had been up the previous night responding to a call for a downed tree. Since then, he had crawled into an attic, saved an entire block of homes, and spent hours defending an antique shop and sporting-goods store. Johnson was among the crews who fought for hours to protect the Optimo, only to have it burn down in front of them.

It was clear to Johnson that Sedwick was quite deaf. Moving closer as they spoke he still had to shout, trying to tell him how dangerous things were, and that he should evacuate.

By 11:30 p.m., the fire had grown to at least 31 square miles. In Paradise it had destroyed thousands of homes, and it was still on a rolling boil there and in Magalia. About fifteen firefighters in front of the Depot, including Johnson, were preparing for what they knew would be an intense fight. All day their primary objective had been to get as many people out as quickly as possible— but that night their assignment was to save the Depot itself.

Sedwick had seen fires in and around the canyon for more than seventy years, and they all tended to do the same thing. Sparking near Jarbo Gap, they would race across the foothills toward the West Branch Feather River. They never jumped the water, yet today they had. Despite the fact that his cherished Ridge was aflame, Sedwick was still confident his tiny corner of Magalia would be safe. "I've seen this a million times over," he told Johnson, looking at the canyon. "It'll be fine."

But the firefighters expected a bad night. The wind was relentless. It shook the 100-foot-tall pines that lined the road back and forth, making the gigantic trees sway like feathers. Johnson, standing in his yellow Nomex suit, now blackened, and red captain's helmet, watched them with an uneasy eye. The fire was still burning in Old Magalia just a few hundred yards southwest, and to the east there was a glow creeping through the canyon that hugs the hill. This was not a blaze the firefighters thought they could predict.

The devil is at our heels, Johnson thought to himself.

He had to get to work. The Depot meant everything to the town. It was where Johnson would come for a beer after a long day of mountain biking. Johnson knew they couldn't let this building burn. It would be like watching San Francisco's Coit Tower collapse.

Firefighters from Magalia's Station 33, the county, and the state's emergency office bustled about. There was no power, and the smoke had blocked the stars. It was dark save for headlights on their engines, glinting trucks some 35 feet long. They were backed into the Depot parking lot so that if things got bad they could flee. Meanwhile Sedwick, Johnson noted, had headed up the driveway toward his home.

The crews started their defense of the Depot by removing boards lining the side of the building, and then the patio furniture and any trash. They wetted the grass and brush around the Depot for about an hour, mindful of the upcanyon wind sending embers in their direction. It fanned the fire in the gorge, gusting it up the canyon walls. Fires burn faster uphill because heat

rises, quickly warming and drying out fuels upslope, and rapidly moving uphill fires have been deadly for firefighters. Twelve
smokejumpers—the elite wildland crews who parachute into isolated blazes—and a fire guard died in the notorious Mann Gulch
Fire in Montana in 1949. On a hot and windy August day, they
had been picking their way down a canyon when, unbeknownst
to them, they had been outflanked by a fire that reached the bot-
tom of the gully before them, and began tearing uphill in their
direction. They ran, but it torched 3,000 acres in ten minutes and
fatally overtook them.

The embers lofting into the sky from the West Branch Feather
River Canyon ignited a gray pine down the street, only a stone's
throw away. The temperature climbed as Johnson and the others
continued hosing down the Depot, soaking it as much as possible.
A thick row of ponderosas that lined the canyon directly across
the road from the restaurant caught fire, sending flames 150 feet
in the air. The wind directed them straight at the Depot. Firebrands fell in the parking lot like hail and the noise of the inferno
drowned out all other sounds.

If threatened by imminent danger, the firefighters had planned
to caravan out to the Magalia Community Church, a third of
a mile to the southwest. Now the wall of flames danced over
them, pushing fire over Depot Lane, the only way out. They had
to move.

Johnson tried to unscrew his hose, but the flames seemed to be
right on top of him. When overrun, crews sometimes cut their
rubber hoses and flee with the valuable brass nozzles in hand, but
because the heat from these flames was so overwhelming, Johnson didn't even have time to get his axe out. He would have to
drive away dragging the 150-foot hose until it snapped.

Fire cascaded down on top of the Depot, the building Sedwick
had lived next to and loved for more than seventy years. The

building the firefighters had, for over an hour and a half, been giving their all to save. Its yellow paint was seared by the heat. Smoke grayed the faded painting of mountains and pines along its foundation. Inside, unseen, the new appliances, stoves, and a wine cooler, which the owners had replaced after a fire two summers ago, would have lost their shape and started to smoke.

As the brush and trees around the buildings ignited, the crews piled into their engines.

According to firefighters, the fire emitted radiant heat reaching 1,500 degrees Fahrenheit, as hot as a crematorium. Trees across the street were burning so hot that they were left looking like charred matchsticks.

Sedwick was somewhere behind the restaurant, up the driveway that led to his cabin. He believed his home would be safe because it always had been. But embers floated onto his property, sparking spot fires and cutting off his escape route. Sedwick would have been boxed in by fire, and directly in the path of the flames barreling toward him.

Ray Johnson knew that Sedwick was on the driveway, and he also knew that there was absolutely no way to reach him. He looked for Sedwick, but his view was obscured by a curtain of flame that closed between them.

Farther along Sedwick's driveway, flames licked the walls and roof of the old orange-walled cabin that he could never bear to change. Gathering strength, they invaded the building and ate through the portrait of Sedwick's mother, the cedar chest she brought with her from Scotland a hundred years before, and the pages of his years-in-the-making novel. The bathtub on the porch grew warped and misshapen. The loft bedroom where Sedwick slept fell into the house. And the enclosed porch collapsed.

In the end, his niece would say, Sedwick and the home he loved so dearly left this earth together.

For twenty hours, Bubba Shipman had, like Sedwick, been fighting to save his Old Magalia neighborhood—the homes next to his, the old church—unaware of how the rest of the Ridge was faring. For the last four, he'd been in the dense forest behind his home, using his chainsaw to cut through the trees and dry brush to create a fire line.

He toiled away, covered in sweat, with a gash on his left eyelid where a sharp piece of charred bark had cut him. A flap of skin hung over his eye. He expected that the wind would shift in the morning and he wanted to get rid of any fuel near his house.

But fire came hours sooner than he expected. Trudging up the steep gravel incline to his house on Magalia Cemetery Road, Shipman could feel the heat at his back, and abandoned the chainsaw. The air felt too hot to breathe. "You gotta move," he told himself.

Trees behind him popped with a screech that sounded demonic. Knees buckling, he fell to the ground in front of the house of a neighbor who had asked Shipman to rescue his dog. Shipman rallied himself and kicked in the door, picked up the Australian shepherd, and carried it out.

Reaching his home next to the old graveyard, he went for his other chainsaw and started cutting down more brush. The chain fell off and a nearby house went up. Shipman got down on his knees in tears: "God, I just worked all day long for everybody else. Now you're going to take mine?"

His neighbor's propane tank, just 10 feet away, started venting, sending streams of flames 50 feet into the air and directly toward his home, and his massive cedar.

It was an elegant, 100-foot tree growing against his house. It offered good shade in the summer. But looking at it now, all he could think was how it would spell the end of the life he'd built. Many of its limbs were dead and brittle, and woodpeckers had drilled holes that dried it out from the inside.

The tree was an unwieldy 4 feet wide at the base, but Shipman hoped he could get it to fall on his carport, away from the cemetery. Embers fell onto his hands as he cut and melted the plastic of the chainsaw. Finally, Shipman heard a snap, and the cedar started to fall.

But it came down the wrong way. Instead of collapsing over his carport, it went east, toward the cemetery, filled with the remains of the Paradise Ridge's earliest settlers. There was the grave of Peter Woolever, a Scotsman who came to California in the 1850s and built a hotel in Magalia that served the gold miners who flocked to the area. Or another that only listed the name Ada, with a dedication to a mother and child who died together in 1892.

The cedar brought down a section of chain link fence and crashed over several of the headstones. There was an almighty bang, and one cracked down the middle.

It was 4:00 a.m. on Friday, November 9, and the destruction of the Paradise Ridge as it had been known for a century and a half was almost complete.

7

The Day After

At daybreak on November 9, when the rising sun hesi-
tatingly reached through the haze and the flames had
abated, Anna Dise got up from her neighbor's gravel
driveway. She had stayed huddled there for around eleven hours
overnight, dousing herself with water as the fire rose and fell.
Although bone-weary, she'd made herself stay awake because she
thought that if she nodded off, it would be the end. Her father was
gone, but this was not a time for grief. She still had to survive,
and perhaps she had found the one protected spot where that was
possible. She'd made up stories to keep her mind busy. Seeming
unafraid of her and her dog in these unearthly circumstances, a
deer and a skunk came to drink from the stream.

She walked to her neighbor's duck pond to drink, long past
the point of worrying whether it was clean. She found her other
dog, Sirius, lying in the pathway on her property, covered in ash.
It took a few attempts to rouse him. Everything else was man-
gled, collapsed, ruined, or burning. She couldn't see anything
that resembled a body or bones.

Back on the road, Anna and the dogs headed toward the valley
and the sound of chainsaws. They crawled underneath a fallen

tree. They were joined by a German shepherd that had also ridden out the dreadful night. By the look on the firefighters' faces, she could have been a ghost.

When Iris Natividad still had no word from her partner, Andrew Downer, by 5:00 p.m. on the day of the fire, she knew logically what must have happened. She'd continued to ring Downer's cellphone throughout the day, hoping he'd pick up. She, his siblings, and his caregiver "all started calling hospitals, shelters, calling any place we could think of that would care for a disabled person," she said. If he wasn't in Paradise, maybe he was in Chico. If not Chico, perhaps Oroville. "I felt like that parent that lost her child, that's how I felt. I felt like how those parents feel that have their child kidnapped and they're agonizing over what, where, how." She stayed in Santa Rosa that night, because people told her not to come back. It was total chaos.

Finally she was able to arrange for a "welfare check." A firefighter visited their home, saw that it was destroyed, and on the porch found Andrew's burned wheelchair and his remains. He had waited there for the firefighters, just as Iris has told him. There was no sign of the dogs.

"Andrew was like the captain of his ship," Iris said. "And he went down with it."

For two days, Lonnie Walker, the truck driver who had been on the road 100 miles away, had no idea what had happened to his wife Ellen. He spent the first night sleeping in his pickup near Oroville because police wouldn't let him inside the cordon they had set up. After that, the trucking company he worked for provided him with clothes to wear and a place to live in Yuba City. Waiting at local hospitals, his hopes would surge when he saw an ambulance pull in. He searched the shelters. He kept asking the police to check his house, but he felt that they were blowing him off. Walker still thought there was a chance he could rescue Ellen, because during the Humboldt Fire in 2008, their neighbors' homes had burned while theirs had been spared.

The pastor of the Walkers' church, Golden Feather Seventh Day Adventist, where Ellen sang, heard of Walker's misery. He contacted Walker's friend, a boat mechanic named Randy Saunders, and the two met in a Denny's parking lot. If Ellen had made it, Saunders thought, she would have found a way to let her husband know. Nevertheless, Saunders told him: *Shoot, Lonnie, I've got a 4-wheel-drive truck. Worst-case scenario, we'll find a way to sneak in there.*

To get to the Walkers' home on Schwyhart Lane in Concow, Saunders first tried a back road that would allow them to cut midway onto the wide, speedy two-lane road, Highway 70, that led into the hills, hopefully dodging any checkpoints lower down. But the street was blocked by a sheriff's car. Saunders spoke to the man through his open window: "*Look here, Lonnie's wife has been missing now for over two days. We just want to go and check the house and see what's going on.*" The policeman said he couldn't let them through and offered to call in a request for an officer to go up there, a response Walker had heard so many times before. Walker gave his name and number and was promised an answer within the hour.

Saunders wanted to try a couple other roads, but the next one was also blockaded by a national guardsman and a highway patrolman. He appealed to the latter: "*Dude, don't you understand, this is this guy's wife. Two days. He says, Well, there's nothing we can do.*"

"I looked at Lonnie and I could see the look on Lonnie's face. I started to back up. And I said, *Do me a favor, don't arrest me till I come out.*"

Saunders floored it. The speedometer nudged 100 mph. In his rear-view mirror, he saw the cop pulling out and following them but not at any great pace, and he thought they might have made it. But it quickly became clear why they were not being chased. Another checkpoint had been installed farther up the highway. The officers blocked the road with their vehicles. "They drew their guns on us and told us to shut the car off, and I did. One guy holding a gun on us said, *Don't move, don't get out of the car, keep your hands where we can see 'em.*"

Behind them the highway patrolman who had been in pursuit pulled up. He was "rather upset at me," as Saunders put it. "He literally pulled me out of the truck, handcuffed me, turned me around. I said, *What I just did is what you guys shoulda done two days ago, and is get up the road and find his wife.*"

He continued: "Don't you realize this is this guy's wife? Two days. What would you do if it was your wife?"

A sergeant showed up, and it just so happened that Saunders had worked on the man's houseboat. "He goes, *Just be cool and stay here and let's see if we can work this out.*"

An officer talked with Walker, who explained how he had called and called and received no help. Two of them agreed to escort him up to his house. The officer who had collared Saunders "undid the handcuffs and spun me back around rather abruptly. He said, *You get in your truck and go back to where you first ran the roadblock, and you keep it under 100 mph.*"

The escapade ended as Saunders thought it might. When Walker got up to his house, he saw a yellow ribbon strung around the rubble of it. They wouldn't let him get close, they said, because it was a crime scene. They took him back down to his friend, who knew what he had seen by the way his head hung low.

"He wanted just to go back to his truck and stay there," Saunders said. "I said, *No, Lonnie, you don't need to be by yourself.* I said, *I know you.* I said, *I just want you to come crash at my place. Leave the truck here. I got a separate room. If you wanna cry, I won't think less of you. I've cried myself before.*"

Walker stayed the night and left. He didn't know what happened to Ellen. He hoped she was asleep. The image of her lying there at 3:00 a.m., when he didn't wake her up, was the most awful, the most treasured photograph in his mind.

"I've always been there for her but this time I wasn't, and it just tears me up because I only had a couple more days and I was going to hang it up," Walker said, referring to his retirement. "I loved that lady more than anything this side of heaven. I'd have given my life for her, not a doubt in my mind. I wish she was here and I was there.

"She hadn't been wanting me to work for the longest time anyway, and I was just too bullheaded to see through it," Walker said. "So I'm paying for it now."

Saunders hated how Walker flagellated himself, but he couldn't talk him out of it.

"*Lonnie, please don't do that,*" he told Walker. "*Don't do that to yourself. You gotta realize—the good Lord—if it was her time to go you couldn't have stopped it.*

"*Things happen for a reason. Someday you'll know about it.*"

Other locals also found ways through the cordon. On the night of the fire, Bill Goggia purchased a flashing amber light like those used by emergency vehicles, put it on his friend's truck, and together they rode back into Paradise without anyone bothering them. The extent of Goggia's injuries had not yet been recognized—he was in terrible pain, and would later be admitted to the hospital for almost two weeks. But he was determined.

Below the crosses, there was his van, still intact. He opened the door to a meow.

Since evacuating without her mother, Chellie Saleen had spent four days living in her car because there was no space in the evacuation centers. "I froze my ass off," she said. She was bitten by a spider. And she did the rounds of shelters, filling out a missing-persons report at each one, and praying to find her mother, Joanne Caddy. About five days after the fire she was told that Caddy's home was gone but that no remains had been found; a couple days later she learned that there was, in fact, a body. "I just felt the center of my stomach drop out," Saleen said. She isn't sure how her mother, or her mother's two dogs, died.

Saleen's own home was destroyed. Sitting outside a shelter in a Chico church on a frigid morning, she was forlorn. "She was probably expecting me," she said. "I hope she was just sedated and sleeping and didn't suffer."

In Chico on the day after the fire, Tom LeBlanc found two voicemails that his stepdaughter Kimber Wehr had left the morning before. *What's going on outside,* Kimber asks in one. *Is this the apocalypse?*

Each day from then on was "fraught with unanswered questions and tragedy," Pat LeBlanc said. She and her husband toyed with the idea that her daughter escaped. "Maybe she got out and disappeared and went to some other town and was so traumatized she wouldn't answer phone calls. That could have been so, given her personality." The couple visited evacuation centers, and around five days after the fire Pat gave a DNA sample. That whole period was a "black nightmare," Pat said.

Pat kept returning to how, just recently, she and her daughter had a profound rapprochement after decades when she felt like she didn't quite understand her. A few weeks before the fire, Pat received an unusual late-night call from Wehr, when they talked for over an hour about Pat's pancreatic cancer. "She told me I couldn't die first, that it was my obligation to stay alive and show her how to navigate seniorhood," Pat said. "I told her I wasn't intending to die, I would beat it somehow." Wehr had a premonition, Pat later thought, though perhaps it was for the wrong person.

Following the phone call, Pat invited Wehr over for a haircut—an occasion when they might celebrate Wehr's passage into seniorhood, Pat joked, even though she thought her 53-year-old daughter still far from elderly. To her surprise, Wehr said yes. Wehr had gotten into trouble when she was young for cutting her own hair, as kids sometimes do, and for unclear reasons this lingered bitterly in her memory. Afterwards, when Pat persuaded her to look in a mirror, something she ordinarily never liked to do, she beamed. "I'd never seen her smile over anything like that, and she was thrilled, really she was. It was a very special moment and had much more to do with things other than hair," Pat said. "She went happily off. She gave me a big hug at the door."

Wehr, her mother belatedly realized, simply did not, and would not, operate according to the same rules as she did. "I'd missed out on a great deal because I'd kept waiting for her to be more like me, and it didn't happen," she said.

Several days after Pat gave her sample, a detective called and asked to visit them. Once the appalling news had been delivered, Wehr's voicemails became poisonous with new significance. "I can't use that cellphone and I can't erase the words," Pat said.

Although there had been no way for her to reach her daughter, she nevertheless blamed herself for not reaching her. "As a parent you feel you must always save your child, and if you can't do that, what use are you as a parent?" she said.

From the road, Christina Taft had called and called but could not get through to her mother, Victoria. The traffic had been a nightmare and there was no turning back even if she had wanted to. After getting down the hill she met with a college friend who drove her from shelter to shelter in search of her mother. She rang emergency hotlines but felt like she wasn't being taken seriously. Her mother's death was finally confirmed over two weeks later.

Victoria's remains were found in their living room. But Christina already knew.

She was agonized. "Twisted," she said. "How I was there and didn't drive her out. She was my best friend. She couldn't drive, she had disabilities." At other times she was furious over how they had been abandoned. "We didn't get a single call or text or anything." She was jealous of those who did.

Blasting music from her car stereo, she drove in circles. "When she passed I felt like there was no one to wonder where I was going," she said. "Or if I disappeared, what had happened to me." After moving into student accommodation at Chico State, she devoted herself to researching her mother's past, uncovering her story from old resumes and film and TV recordings, and speaking with her old friends. "It helped that she had a real good life," Taft said; one boyfriend called her mother the "golden girl."

Like a hateful broken record, the choices that she and her mother made during those last hours echoed in her mind, along with the tantalizing ghosts of the alternative paths they could have taken.

"I go in circles of blame around what was said, the factors in the months leading up to it," Taft said. "It feels like we were careening off a cliff. We didn't know a cliff was there and we were careening off."

The day after the fire, Angela Loo had still not had any news about her father, Ernie Foss, or his stepson, Andrew Burt, in Paradise. She was teaching a pharmacy class at Portland Community College when she got a call from Butte County. She knew what it would say, anticipated the black hole that was about to swallow her, so she finished the class before returning it.

Foss's remains had been found with his wheelchair in the driveway. As best as Loo can piece together, Burt managed to get Foss into his chair and out of the house, but it was too much to get him from the chair into the minivan. It looked like the hosepipe had been pulled out, perhaps for a last-ditch firefighting effort. But water pressure had failed across town.

That left Burt, who was missing. Loo suspected that Foss knew he would be unable to escape, and that it would have been hard for Burt to make a run for it with him in tow. She thought he would have instructed Burt to go without him.

A week later, Loo learned that Burt was one of the bodies in the terrible video shot by Greg Woodcox. The remains of Bernice, the devoted dog, were on the asphalt next to him.

On the morning of November 9, under a beige sky, Skye Sedwick decided she couldn't stand to breathe Butte County air a second longer.

Pooling their cash, Skye and her boyfriend headed north to a town called Anderson and got a basic room at a Motel 6. They could only afford it for one night, but Skye was grateful when she woke to blue skies the next morning.

Her father had stopped responding to calls or messages from her or anyone. It was completely unlike him.

Days passed in a cold, exhausting blur. As they waited for answers, she and Merritt picked up donations of food and clothes, and one afternoon a kind woman handed Skye a cup of hot chocolate. It was, Skye thought, the best she'd ever had. Bouncing from place to place, they slept one night in Merritt's truck in the parking lot of a twenty-four-hour grocery store. When they went inside to shop, some of the donations they received were stolen from the vehicle. Eventually a friend of a friend of a friend invited them to stay at his house 50 miles south, near Marysville. It was filthy, covered in cobwebs so thick they looked like Halloween decorations. Cleaning it took days.

With every day that passed without word from her father, Skye grew more certain he was gone. Looking through her phone, Sedwick's niece, Simona MacAngus, discovered a partly unintelligible voicemail from him that she had missed, and that he had left her late on the night of the fire.

"Paradise . . . thousands of houses gone. Pray. The folks around here, I don't know where the hell they're gonna live. Sawmill Peak is on fire, been on fire all day. The winds blowing, that's what caused it. And everything is dry. Of course, they took. . . . I tried several houses, to save several houses. I couldn't do it. I did save the firehouse. The people. . . . and they're kinda poor people but they fought the fire all day. Anyway, I'll. . . . God bless you."

PART THREE
ASHES
AND SEEDS

8

A City Dispersed

The fire that destroyed Paradise arrived in San Francisco that same Thursday afternoon in the form of a piney smell reminiscent of a campfire. It was followed by a thick blanket of smoke. Most people didn't know that the town it came from was already nearly gone. Unlike San Francisco's acrobatic, off-white fog that flowed from neighborhood to neighborhood, this was a deadening sepia mantle draped over the entire city, from the Golden Gate Bridge to the Ferry Building and the Mission district. Office workers and tourists still streamed among the skyscrapers downtown and along the Embarcadero, but as the smoke lingered many started to wear white masks, making it appear as if they feared an airborne pathogen. Authorities recommended n95 respirators, so called because they claimed to filter out at least 95 percent of dust and mold in the air. The masks became coveted items as the air quality worsened, and hardware stores sold out of them almost as quickly as new shipments arrived.

John Sedwick was born in San Francisco, and his family had moved to the Paradise Ridge from the Bay Area more than seventy years earlier so the asthmatic boy could breathe with ease. Now his adopted home was suffocating his birthplace. As winds blew smoke from the fire south, the air quality index, which measures dangerous smoke and particulates in the air, would reach a level deemed "very unhealthy," putting everyone in the population at risk of respiratory problems. Conditions were dire north of San Francisco, in Sacramento and Chico, where the air quality became the worst in the world, more dangerous than in New Delhi or Beijing. An inversion layer, in which warm air rises over cool air and creates an artificial ceiling, trapped the smoke, and those immersed in it inhaled the equivalent of ten or so cigarettes' worth of harmful particles a day. Gradually it filled the great bowl of the Central Valley, from Paradise to San Francisco, and it funneled out to sea through the San Francisco Bay Area.

Coffee shops in the city were full while parks were empty, though the temperatures were in the 60s. Parents wouldn't let their kids play outdoors. The San Francisco Unified School District canceled school for its approximately 57,000 students. The city shut down its famous cable cars. UC Berkeley and Stanford postponed their annual "Big Game." Volunteers distributed masks to homeless people on the streets.

Before European settlement, California may have spent much of the summer and fall cloaked in smoke like this, as it smoldered from lightning strikes and burns started by Native Americans. With tens of millions more people living in the state, worsening wildfire "smoke waves," defined as a period when dangerous particulate matter from a blaze has smothered an area for two or more days, were being called a public health crisis. A study found that smoke-related deaths in the United States could increase to 40,000 per year by the end of the century, more than double the number in 2018. One outdoors magazine even suggested that hikers planning a trip out West travel by July at the latest, to avoid the months with the highest risk of smoke and flame.

While the Camp Fire ravaged Butte County to the north, a

blaze called the Woolsey Fire was working its way through Los Angeles and Ventura counties and would end up torching 97,000 acres, destroying homes and historic movie sets, and scorching swathes of the Santa Monica Mountains. It killed three, including a mother and son who died trying to flee in their car. Their bodies were later found in the charred vehicle. Smoke from that fire largely drifted out over the Pacific, though it towered over the region like a volcanic eruption.

The impacts of the Camp Fire radiated outward from its ignition point. Fifteen miles down the road from Paradise, a few days after the fire started, a cattle rancher named Megan Brown trudged through the mud toward an excited drift of pigs that waddled in her direction. She reached down and petted a large brown hog on the head. "This is Kevin Bacon," she said affectionately, her words slightly muffled behind her disposable mask. "He's a good boy." She grabbed five-gallon buckets, filled them with grain, and poured them in the feeder.

Brown was a sixth-generation rancher in fire country. The smoke at Table Mountain Ranch brought back memories of the year before, when a wildfire had burned onto her land and caused $4 million worth of damage, destroying several buildings, including two homes. Technically, Brown was under an evacuation order. She and her mother had fled once when they saw flames and returned after firefighters got things under control. Still, they remained packed and prepared with family photos and clothes in the backseat of the truck. They slept in shifts to keep an eye on the fire, but couldn't bring themselves to leave. That would mean abandoning the animals: eighty pair of cattle, twenty chickens, and forty pigs along with a litter of piglets. Though they had moved their horses off-site, they didn't have the resources to relocate the other creatures. She became emotional at the thought of what might befall them. "If all these animals die . . . this is my passion, my life. It's my legacy. I would lay down my life for this place," Brown said as Kevin licked her shoe. She could tell the smoke was getting to them. The animals seemed lethargic and had trouble breathing. Weeks later, she'd find a sow cold and lifeless.

Brown stayed and worked, and by the end of each day her white mask turned a sooty gray. She knew that the smoke wasn't just a harbinger of a town burning in the distance. It *was* the town. It was composed of all the trees and the homes that burned, the chemicals from the incinerated plastic, insulation, and paint. It was made up of the people who died, too.

I'm breathing them in right now, she thought to herself.

More than 50,000 people had been displaced because of the fire, and most converged on Chico, the city of 90,000 just twenty minutes from Paradise. Immediately the Walmart in south Chico became a gathering place for evacuees from the ridge—an easy-to-find spot where they could meet once they made it down the hill—and it quickly morphed into a refugee camp. Dozens stayed on in their cars or slept outside. There were RVs, pickup trucks, and vans piled high with whatever people could throw in before they fled: toys, pillows, family albums.

Andrew Duran, who had tried, unsuccessfully, to save his mother's home in Paradise, was one of the many bedding down in a sleeping bag in a field of yellowed grass next to the big-box store. Tents had sprung up all around him. The smoke overhead had blocked out the sun and plunged the town into an unnatural chill. It was hard to breathe, but Duran didn't wear a mask because he doubted it would do any good.

Duran made friends with two other evacuees, Tammy Mezera and Daryl Merritt, and from the second night on they let him sleep under the side canopy of their tent. They were out of room inside after offering a spot to a neighbor who had been sleeping under a taco truck. "We are just waiting," Mezera said. "Waiting for information. Waiting for something so we can get back to some kind of normalcy."

Lodging in the town, from the economical Safari Inn to the upscale Hotel Diamond, quickly filled up. Many evacuees stayed with family and friends. About 1,400 went to shelters at churches, fairgrounds, and a school, as far as 75 miles away. Others found empty lots or driveways where they could park their RVs and trailers and live in them.

Hospitals took in transfers from the Ridge. When Rachelle Sanders reached the Enloe Medical Center in Chico after her eight-hour evacuation from Feather River hospital, only twenty hours after giving birth, she was naked, because her flimsy hospital gown wouldn't stay on as she tried to nurse Lincoln, and she still had a catheter inserted. The hospital was expecting her, and had multiple staffers waiting for her outside a side entrance along a tree-lined street.

Sanders had been trapped in Paradise for an age, stuck in traffic and directed in circles. She hadn't known whether she could save herself and baby Lincoln, or if she would ever see her two older children or her husband again. But she and Chris had been reunited and here she was with Lincoln, safe at last. She had kept her composure for most of the day, but in this moment, she couldn't any longer. As nurses wrapped her in a pink robe, placed her in a wheelchair, and took care of Lincoln, Sanders started to sob. She looked at him, swaddled in a blanket, with Chris trailing behind, and couldn't stop. The tears that she hadn't cried came out all at once. They continued for twenty minutes after she had made it to her room, and only finally slowed when the nurses told her that Lincoln looked good—that he was okay. They both were.

Amid the panic and hazy confusion, a terrifying picture was

emerging. "The town is devastated, everything is destroyed," a spokesman for Cal Fire was quoted as saying. "There's nothing much left standing." Lieutenant Governor Gavin Newsom had declared a state of emergency in Butte County before 3:00 p.m. on the day of the fire, authorizing state employees and the office of emergency services to provide assistance to the county. It was a responsibility that would have fallen to Governor Jerry Brown, but with Brown still out of state after a speaking engagement, Newsom was in charge, and he was briefed by Cal Fire about the fast-spreading blaze.

The first night of fear, anxiety, and unknowingness shaded unbearably into another and then the next for the evacuees. They were weary and sickened. In the coming days about 145 people in shelters caught norovirus, a highly contagious illness causing vomiting and diarrhea. One evacuee from Magalia, Maureen Rutty, was spending nights sleeping in her car at Walmart with her four rescued Pomeranians. She worried that if she went to a shelter and caught the virus, she'd be unable to pay hospital bills and would have to be separated from her animals. "If I got sick and had to go to the hospital, and my dogs went to the pound, we're finished," she said in her distinctive accent, a blend of England and Oklahoma. "I can't lose my dogs."

Among evacuees, the need was great, because many people had left with next to nothing. Retiree Susan Van Horn escaped with her "go-bag," some changes of clothing and utility bills. She considered herself lucky. A high-school teacher, Virginia Partain, only had time to pick up her students' draft college admission essays, her cats, and a blanket. Former nurse Marissa Nypl grabbed clothes, a heating pad, and medication, but not the family albums from the living room, or the crystals her husband had collected on a trip to Mount Erebus in Antarctica. Retired teacher Suzanne Linebarger left with $30 worth of scallops she had purchased for dinner, a pair of mismatched shoes, and her swimming fins from a vacation to Mexico. "Why didn't I grab my grandmother's mandolin worth hundreds of dollars?" she wondered.

Food-truck owners like Jose Uriarte stepped into the breach. Uriarte owned Gordo Burrito in Chico and drove his truck to the Walmart, where he handed tacos, tamales, and burritos to shell-shocked evacuees and first responders. Church groups from around the state cooked and distributed meals. People sent and left thousands of pounds of donations: clothes, food, and household goods, until officials had to ask people to stop because shelters and drop-off locations had reached capacity. Cash and gift cards flowed in from across the country. Aaron Rodgers, an NFL star with roots in Chico, donated $1 million to launch the Butte Strong fund and assist with the rebuild. Sierra Nevada Brewing contributed $100,000 to it. A Southern California businessman gave a $1,000 check to every student and staff member at Paradise High School. The man who had helped evacuate dozens of seniors from the Apple Tree Village Mobile Home Park in Paradise, Stephen Murray, wrangled donations of trailers and personally drove evacuees to new homes across the United States. Heather Lofholm, a court reporter from the Sacramento area, and her husband Eric, a motivational speaker, created a Facebook group where people who wanted to help were matched with survivors in need of assistance. Within weeks, they had helped to distribute thousands of dollars.

Hundreds of people appealed to strangers on crowdfunding websites. Christina Taft asked for help to cover cremation and funeral costs, the university housing she had been forced to move into, and gas and groceries. And she wrote a simple, devastating tribute to her mother Victoria. "She loved the community, had volunteered in it, and did not want to leave it. I wish we had," Christina said. "I feel pain, guilt, shame, loss, death, and anger about what happened. I wish I had stayed longer and that the emergency response was better." She raised over $8,000.

Official accounts of what had transpired in Paradise emerged slowly. On the day of the fire, authorities announced that it had burned up to a thousand buildings and killed multiple people, but details were scant. At another briefing the following evening— they became a daily event—county sheriff Kory Honea, standing before dozens of reporters, said the fire had left nine people dead. Four sets of remains were found in charred vehicles, and another was discovered a few feet away from one. Authorities encountered victims near or inside burned homes.

By Saturday, authorities had recovered twenty-three bodies from cars and homes in Paradise and Concow. "This event was the worst-case scenario. It's the event we have feared for a long time," Sheriff Honea said, as thick smoke wafted into the room at the Chico fairgrounds. Some people wore their masks even inside. The fire was unlike anything the region had ever seen, Honea said. "You get the sense that it's raining hell down upon you."

By Monday, the death toll had climbed to forty-two, making the Camp Fire the deadliest in state history, topping the Griffith Park Fire, which killed at least twenty-nine people in Los Angeles in 1933. That evening, Honea announced the names of the first identified victims: Ernest Foss of Paradise, Jesus Fernandez of Concow, and Carl Wiley of Magalia.

Outside the shelters, anxious evacuees and out-of-towners hung missing-person posters, looking for relatives and friends they hadn't heard from since the fire or before. As officials combed through 911 calls and a backlog of messages, the number of missing jumped to more than 600, and then to over 1,000. The veracity of these estimates was unclear. Some people were on the list several times with various spellings of their names, because mul-

tiple friends and family members called in. Others hadn't made
contact with family or friends and didn't realize they were considered missing in the first place.

One poster showed Julian Binstock, eighty-eight, and his black-
and-white border collie, Jack. "Please help us find him," read the
caption above a picture of a man with a subtle smile. Binstock's
death was announced nearly three weeks later. His remains were
identified in the rubble of his bungalow at the Feather Canyon
Retirement Community, where he was the only resident not to
be evacuated. Jack was by his side. Family members told reporters they weren't sure why Binstock stayed, but believed he had
been asleep and that staff had tried unsuccessfully to rouse him
until they were forced to flee as flames closed in. Another poster
showed sixty-four-year-old Sheila Santos bundled up in winter
clothes in front of a huge Christmas tree. Santos's daughter, Tammie Konicki, drove from Independence, Ohio, less than twenty-
four hours after the fire started to find her mother. On November
30, she received a call that her mother's remains, found in her
home at Holly Hills Mobile Estates, had been identified.

Families were forced to do the rounds of shelters in search of
loved ones. The relatives of Barbara Carlson had not heard from
her since the morning of the fire, when she told them she wasn't
planning to leave the home she shared with her sister, Shirley
Haley, on Heavenly Place. Carlson was a quiet book lover with
three children and seven grandchildren who moved to Paradise
after her husband of more than thirty years died, and Haley was
a devoutly religious woman who came to the area in the 1990s
following a career in the medical field.

Carlson's son Mike and granddaughter Annika drove more
than two hours to shelters in Oroville, Gridley, Live Oak, and
Chico, looking for the small woman with gray hair and freck-
les and a dog called Strawberry. "We are praying she's with her

neighbors," Annika said in between shelter stops. "My grandma is a sweet, kind old lady with a huge heart but she is probably in a lot of distress and very emotional because she has lost everything."

Three days later, Annika posted an update on Twitter. "Thank you for all the amazing support from everyone. We got confirmation the house where my grandma and her sister lived had two bodies found today. Barbara Carlson and Shirley Haley were two loved victims of the Camp Fire."

The family didn't know why they had stayed, or if they had been unable to leave. Both women had been in good health, with access to a car and cellphones, and they'd known about the fire.

A family friend had in fact called Haley to tell her to evacuate, Annika said, but Haley had replied that, according to God, "it would be safe to stay." Still, the family knew that firefighters had driven near Haley and Carlson's street with sirens and warned people to leave. "Maybe they tried to and the car didn't start? We don't know," Annika said.

It was incomprehensible just how swiftly an entire world had been lost. It wasn't only this home or that. It was also the Ridge's history, in the form of the 50,000 or so items housed at the Gold Nugget museum, from its complete antique blacksmith shop to its doll collection, printing press, and gold-panning equipment. One local couple had published a meticulous guidebook to the hiking trails and old mining flumes of the Paradise Ridge, where they had lived for thirty-two years, only to see the landscape rendered unrecognizable. Soon after, they moved to Oregon. "We're almost hesitant to even go back to look because it's going to break our hearts," said coauthor Roger Ekins.

A woman described on Facebook the moment she realized she had fled her home without an inestimably precious item she referred to as "Gabriel's box": "His box contained everything I had from his little life, his blanket, his little outfit, his beanie I would often hold in my darkest moments, a poem from the hospital, his footprints." Another resident wrote of her young son and his best friend, who now had to live hours apart. "They giggled and belly laughed often about who knows what—it's a type of giggle I've only ever heard my son do with him," she said. "This is our biggest loss."

A medical technologist called Trisha Wells couldn't shake the sickening, trapped-in-quicksand feeling of a bad dream. Born and raised in Paradise, like her mother and grandfather, she worked at the hospital and also ran Enchantment Children's Parties, which provided performers who would dress up as famous kids' characters. She had helped evacuate the hospital on the day of the fire, and that was the last time she saw it or her colleagues.

"Even though my family is OK I just feel like I can't believe it's all gone," Wells said. "And I couldn't give a crap about my house—I don't care about my bed, TV, computer, any of that stupid stuff. It's like every day you think of something you cannot replace. My husband's grandmother painted this stupid lamp, and we called it the Ugly Lamp, because it was hideous. But she's gone and now it's gone and we can't get it back."

Wells and her husband and their three kids stayed in a trailer for three weeks, until her husband's best friend offered them a place to stay in Virginia, which was where they remained. Her new license plate read PDSE4VR.

With Paradise cordoned off and nearly everyone evacuated, officials ran the town in exile. A few days after the fire, Dave Sullivan, a senior staffer at the department of emergency management in the city of San Francisco, volunteered to go to Paradise and assist with the recovery operations. Several colleagues joined him, and over the following weeks dozens of officials from cities across the state also signed up as part of California's mutual aid system.

In a wildfire smoke–filled room at Chico's Fire Training Center, using their personal laptops, Sullivan and the team helped staff the Paradise Emergency Operations Center, which city manager Lauren Gill had launched on the day of the fire before evacuating. Their responsibilities were all-encompassing, from reestablishing telecommunications in town to equipping public-works employees with chainsaws to clear trees and ensuring that radioactive pharmaceuticals and equipment used at the hospital were not compromised.

For several days, hundreds of displaced residents filed into Cal State Chico's Laxson Auditorium, a 1,300-person venue that had hosted speakers such as Mikhail Gorbachev and Desmond Tutu, and took a seat on the red-cushioned chairs for updates from authorities. The meetings were long and often emotional as the county sheriff, town government, police chief, Cal Fire officials, and a PG&E vice president took turns standing on the stage, providing updates and answering questions from traumatized Ridge residents.

"I want to assure you that the town is working very hard with the state and federal government for all the steps required to get you all back into town," the mayor, Jody Jones, said.

Gill, the city manager, teared up as she offered her condolences: "I just want to make sure that we're very sensitive and respectful to those who have not only lost their business and their homes but . . . the lives that have been lost is going to be tremendous," she said. "I've always said that wasn't going to happen"—she paused briefly—"on my watch." Sheriff Honea, his voice cracking, carefully described how his office had formed a team to recover bodies, and he outlined the process of identifying remains from a fire that burned so hot and ferociously it left only fragments. "You are my people. I am the sheriff of everybody in Butte County including the residents of the Ridge. And you know my special connection to your department. I am proud to be your sheriff," he told residents.

Someone from the crowd yelled "We're proud to have you as our sheriff!" to thunderous applause. "I wish that we weren't doing this," the sheriff said, "but if we have to go through it I wanna go through it with you."

During one of Q&A sessions, Robert Edwards, who had driven from Seattle the night before to find his mother, Barbara Allen, asked if any of the hundreds of evacuees had seen her. His mother's phone had been stolen the previous week, and he had sent her another that was scheduled to arrive the day of the fire. When Allen didn't make contact, and calls to her landline didn't go through, Edwards decided on Friday to come and look for her. He arrived in the area early Saturday. The fifty-two-year-old, a friend who had journeyed with him, and Edwards's cousin spent the day searching shelters all around the county, and walking the Walmart passing out missing-person posters.

Edwards had an advantage that many people didn't; his mother, a petite, lively woman, was well known around town and in the community, thanks to her work with the senior center and the Veterans of Foreign Wars association. At least one person said they knew Allen had escaped from Paradise. It gave Edwards hope, but as the day passed and they didn't find her, his unease grew. He started to think maybe she hadn't made it out. He went to the meeting with the idea that he might spot her face in the crowd, and when she didn't appear, Edwards asked those in the auditorium to help him.

"I just want her to call," he said.

Someone from the crowd shouted: "We've seen her today!"

"Where?" Edwards asked.

"Walmart!" The man said.

The room broke out in applause. Still holding a missing-person poster with the blond seventy-seven-year-old's smiling face, Edwards and his family left the meeting and walked to the lobby. It wasn't much to go on, but they were optimistic. Though it was after 8:00 p.m., they decided to return to the Walmart. They had already knocked on the door of every trailer, but they resolved to try each one again until they found her.

Edwards circled past the food trucks, barbecues, and donation tables. About ten minutes after he arrived, as he did what felt like his hundredth circuit that day, his cousin, also searching, called and said: "I found her!"

Edwards put his foot on the gas and rushed to the old white trailer his cousin had indicated. He recognized it—they had knocked on that door earlier, but no one had answered. He was still pulling the keys out of the ignition as he jumped from the truck. The glow of the tall parking lot lights illuminated Allen's face as she stood waiting in the doorway, her arms open. Edwards strode toward her purposefully, not saying a word. "I didn't have a phone," Allen cried. "I had no way of getting ahold of you." Edwards just hugged her tight and didn't let go.

Joe Kennedy, the firefighter who saved the evacuating nurses and officers, as well as the trapped fire truck, returned home from Paradise for good after ten days, exhausted. He lived near Grass Valley, a quaint and tourist-friendly town in the hills about 70 miles south of Paradise. In and among Gold Rush–era buildings there were wineries, tattoo parlors, and vintage-clothing stores. Kennedy didn't like adulation or social media, though he had received a new nickname. After one of the women he saved called him her "angel," he became known by his colleagues as Iron Angel. It was impossible for him to shake the image of the bodies that he saw through his side window as he pushed the cars out of the way.

"Every fire season has an effect on you and changes you," he said. "I guess I value family a lot more since I saw so many people lose theirs."

The Camp Fire had shown him another thing: he was at risk. The same kind of fire could hit the town he lived in, or any of the other hundreds of towns dotting the forests and chaparral of the American West. Their numbers had already been called.

He mentioned his small community just a few minutes down the road.

"When it burns, they're going to write a book about it," he said.

The casual way he used the word "when," rather than "if," hung in the air.

At his fire station nearby, where the machinery sat mute and buffed to a high shine, he added that he'd probably be on duty at the time it happened. So he'd given his partner these instructions: "Everything in the house is replaceable in one way or another. Just get our daughter and dog and go."

The day after the fire, PG&E had announced to the public that there had been "no determination on the causes of the Camp Fire," and that it would "fully cooperate with any investigations." It had, however, already provided an incident report to the California utilities regulator. That report stated that an aerial patrol had observed damage on a tower on the Caribou-Palermo transmission line.

Investigators from Cal Fire initially prevented PG&E from physically accessing the tower, though almost a week after the blaze began they allowed the company in to help them collect evidence. In its retelling of what it found, PG&E said it saw a broken C-hook—a piece of equipment made of thick, heavy steel and shaped like a "C." It appeared, in layman's terms, that a hook on the tower had snapped and allowed a wire to fall. The company said it had observed a flash mark on the tower, with the implication that the wire that fell when the hook broke hit the tower as it dropped and threw sparks.

"The cause of these incidents has not been determined," PG&E concluded, "and may not be fully understood until additional information becomes available."

The public meetings at Cal State Chico soon became the setting for a nascent and fraught reckoning. At one gathering, a Magalia resident asked why he hadn't received an alert about the fire the morning it started. The sheriff told him that the fire moved so quickly it was difficult to determine where it was headed before notifying residents. The sheriff added that he would probably never be able to offer an answer that would satisfy the man.

At another meeting, PG&E vice president of electrical operations Aaron Johnson, a former energy advisor to the state public utilities commission, briefed evacuees on the company's efforts to clear downed power lines and eventually restore service. "You are our customers, our neighbors, and our friends," he said, gripping the edges of the podium. "We know that thousands of homes and businesses have been lost. That includes the homes of at least 61 of our own employees, who live and work in this community." Eight hundred employees were in the area assessing the damage, and the utility had established a base camp capable of housing 1,500 workers at the Tuscan Ridge golf course, just south of Paradise. Electricity was still turned off.

Johnson was about three minutes in when a man stood and walked toward the stage and asked for answers on a different topic.

"Are you guys responsible," he said, pausing. "For the fire?"

Johnson looked down. And the audience started to boo—not at the PG&E executive but at one of their own. *Let the man talk*, the distraught evacuees said to him. *They're trying to get power back to our town.* Johnson was silent as the scene unfolded. He fiddled with the microphone.

Mayor Jones raced to the edge of the stage from where she sat behind Johnson. "Please do not call out questions. We'll have time for questions and answers. Please be respectful to our speakers."

Johnson said that he was finished.

"Please," the man standing near the stage said. "Before you go."

Again, Ridge residents booed. *Sit down*, they yelled. This time, town manager Gill, wearing an oversized "I heart Paradise" shirt, stood up. "We are not accepting any questions right now," she said. "This is not how we conduct meetings in the town," she said. At regular town council meetings, there are scheduled times for public comments.

"Do we have security please," Gill said. "Can we have security ask this person to leave?"

Just leave, the others told him.

Sir, he said to Johnson once more, pleading with him.

After a few seconds, when it started to quiet down, the PG&E vice president said, "I just want to close by saying . . . we have obviously ceased all billing operations and we'll address those matters at the appropriate time. I will be here and happy to answer your questions when we get to that portion of the session."

The audience applauded. For the people sitting in the auditorium, every aspect of normal life had been stripped away. They didn't know whether their houses still stood, or what was left of their town. The fate of the Ridge was out of their hands, but what they could control was how they responded. And they wanted to maintain some semblance of order, even if the world around them was up in flames.

9

Search and Recovery

W hen search-and-recovery teams first entered Paradise, they were taken aback. Their members had been deployed to disasters such as 9/11 and the disintegration and crash of the *Columbia* space shuttle, but this was like nothing they had seen before. "This was miles and miles and miles of homes just gone," said Tim Houweling, the handler of a dog trained to locate cadavers. A former investment banker who worked in New York during the September 11 attacks, Houweling felt a sense of uselessness that prompted a career change. Now he was part of an urban search-and-rescue task force based in the Bay Area town of East Palo Alto, one of twenty-eight such teams operating under the Federal Emergency Management Agency (FEMA) around the United States. In Paradise, "from the beginning we knew we were not going to be finding anybody alive," Houweling said.

His group of forty-five or so specialists, including doctors, engineers, and firefighters, arrived almost a week after the fire. They proceeded from property to property and used a truck-mounted crane to lift the roofs off burnt-out mobile homes, allowing dog teams to search the debris for human remains. Often they encountered mounted antlers and domestic pets instead. It was impossible for the dogs to search every home in their quadrant, so the teams focused on those where people were reported missing or where there was a car in the driveway.

The Menlo Park team also brought in drones to create an aerial mosaic of the fire damage, and to scout more inaccessible areas. About a mile and a half down an unpaved, unmarked road, the operator of one located a burned trailer. There was a body. "It looked like somebody coming out of their home and they were overwhelmed with smoke and flames," Houweling said. People who were discovered outside tended to be in more identifiable form than those who had been inside. Even so, Houweling and his Belgian Malinois and the other handler went in to confirm the find. "We used it as a way to verify that the dogs are working and smelling the right scent—burned human," he said. He let the dog off leash and tried, as always, to stay impassive, so as not to prompt him. The dog gave the alert he had been trained to in the presence of a decomposing corpse: he barked.

There were other kinds of human remains that dogs searched for. After the 2017 Tubbs fire in Santa Rosa, a handler named Lynne Engelbert, who had once been a rescue specialist at NASA, heard of a man searching for the previously cremated remains—known as cremains—of his parents in his burned home, which had been reduced to eight inches of ash. Her dog was able to lead them to a spot where, burrowing down, Engelbert found a distinctive pile of ash containing tiny, broken fragments. "I had forgotten to take a trowel with me, or I didn't know to, so we collected 'Mom and Dad' using a tuna-fish can that his wife found in a melted garbage can in front of the home," she said. "That man and his family had been totally devastated at the thought his mother and father were going to end up in a toxic dump." Instead Engelbert gave him closure. As a principal on a newly created cremains recovery team, she came to Paradise to do the same thing.

Based on Engelbert's estimation, about 10 percent of households on the Paradise Ridge and in other communities affected by recent California fires contained cremains. After the Camp Fire, her team was asked to search for 252 sets. Engelbert and her border collie spent up to 2.5 hours at each of their assigned sites. In the end, the effort yielded 214 cremains, an 85 percent recovery rate. Most of the urns did not survive. Sometimes they were able to find the little metal identification tags placed on bodies before they are incinerated. Or perhaps a dog indicated it had found a scent, but archaeologists couldn't differentiate the ashes from what a wall or a piece of furniture might leave behind. In those cases, she told relatives, "where the dog alerts—the essence of your loved one is there. So if you like, we can collect right where the dog is alerting." She added, "And that's all they want."

Search teams in Paradise learned that most of the remains of people who died in the fire would not be identifiable in the usual ways by coroners—by fingerprints and dental records—as they had essentially been cremated in the extreme temperatures. On the Monday after the fire, a former pediatric ER resident named Richard Selden, the founder of ANDE, a Massachusetts company that specializes in rapid DNA identification, wrote up a three-page proposal for Butte County officials. A few days later, Selden was in the air on his way to California.

Traditional methods of DNA identification can take months, because of the finicky and procedurally laborious nature of the work. ANDE's technology promised a leap forward: an instrument that automated the entire thing, and into which an investigator could insert a forensic sample and get a DNA readout in just two hours. The human genome is composed of about 3 billion base pairs—meaning 3 billion pairs of molecules designated by the letters A, C, G, and T—that are the blueprint for the creation of a living organism. ANDE could identify individuals even if the sample was degraded and was as tiny as five hundred of these molecules, by comparing them with DNA swabs from relatives. The closer the relative, the better the match should be. Selden's company worked with the Department of Homeland Security, the military, and the FBI, but he thought the next frontier for his technology was the identification of disaster victims. In the case of the Camp Fire, it wasn't a commercial prospect for the company—Selden provided ANDE's services on a humanitarian basis, at no charge.

On the plane, Selden received several photographs of what he would be dealing with. "All of a sudden, some of these pictures popped up, and my heart sunk," he said. They showed severely damaged bone. "I learned in many places that for the body bag they had a paper bag of basically charred bone." Burning is one of the most effective ways to destroy DNA. Although ANDE had tested its technology on bones subject to controlled burns, those from the Camp Fire were far more degraded. Selden thought the chance of getting a DNA identification from such samples was about 5 percent.

It was a luminously clear night outside his window, but as the aircraft descended over central California, bad weather closed in. It took a few minutes for Selden to realize that these were not clouds. It was smoke from the Camp Fire. When the plane pulled into the gate, the other aircraft were mere outlines. To Selden, it now felt real.

For twenty-seven days after the fire, anyone who elected to stay on the Paradise Ridge and surrounding areas was stranded there. The police set up cordons on the roads leading up, and few beyond emergency and response crews were allowed in. People

inside found themselves regularly pulled over by the cops, sometimes with approaching officers' hands on their holsters. Those
who remained dared not cross the line back out for supplies or to
see loved ones in case they weren't let back in.

The restrictions were motivated by concerns about looters and
looky-loos. At one partially burned Paradise home, interlopers
had tossed a family's belongings and left a message graffitied on
the wall: "You are thru owners." Below it was a misspelled sig-
nature: "Lottors."

In Concow, a few members of the Carlin family remained at
their property after their swim across the lake, as did the Moaks,
who had discovered them. They fired up their generators and
prepared to live off supplies in their pantries. Another was Jeff
Evans, the owner-operator of a pest control company whose parents shared his home. With his bushy mustache and checked shirt
tucked into his jeans, Evans looked like he'd stepped straight out
of the Gold Rush. For about six hours after the fire broke out he
had toiled alongside his ninety-one-year-old father to chop down
brush and tamp out embers. When he'd finally felt they could
take a breather, he'd ventured out to see how the community had
fared, and made a startling discovery.

Flames still smoldered in the surrounding vegetation as he took
the main road along the rim of the reservoir, but his way was
blocked by five cars parked higgledy-piggledy on the concrete
crossing over the stream. Their owners were nowhere to be seen.
Surrounding bushes were on fire, and he took them down with
a chainsaw that he'd thrown in the back. If first responders were
going to get past, he thought, he'd have to shift the cars out the
way. Perhaps the vanished drivers had left the keys inside.

The sides of the cars had melted and the doors were too hot to
touch, so he grabbed a leather glove his father had left in his truck
for gathering firewood. First he looked inside a white Oldsmobile
Cutlass. No keys. But a movement caught his eye.

There were two puppies on the backseat, one about six months old, the other about a year. Both tried to hide under the front seats, and one dove through his legs and disappeared into the fire. It took a full two minutes to unwedge the squirming younger dog. Confused by this turn of events, Evans carried the pup back to his truck. Next he checked a black Cadillac SUV. He discovered a pug missing two front teeth. When he scooped her up, she anxiously began to bite the air but quieted as he cradled her. Dropping her off at his truck, she and the puppy kept their distance and sat at opposite ends of the cab.

There were keys in the Cadillac, and he moved it to the side of the road and nudged his own pickup forward. A young German shepherd lay in the road next to the car in front, right where the crossing began. Evans tried to tempt her with some biscuits, but she was jumpy and ran in circles around the vehicle whenever Evans tried to grab her. He moved on, and after the crossing, by some mailboxes, he found a blue Dodge truck. He discovered six more dogs there, big tubs of pot in the back, and absurdly large bags of it on the seats. Evans got most of the dogs loaded into his car, and left the doors of the Dodge open so the dogs he left behind would have a familiar place to return to.

It was like coming upon the *Mary Celeste*, the nineteenth-century American ship famously found adrift in the Atlantic with its crew and passengers missing but their cargo and personal effects left behind. Except the mystery made Evans angry: *What kinds of assholes would leave their dogs?*

Arriving back at his home, a beige, ranch-style place with a large workshop and little hexagonal addition where his father kept his ham radios, the five dogs ran inside to meet his own three in a frenzy of tail-wagging. His father, a practical Navy veteran who served during World War II, was standing in his kitchen and watched, bemused, with a half-eaten peanut butter sandwich in his hand.

"What the fuck are we going to do with all these dogs?" he said.

Evans replied that he couldn't just leave them where he found them.

"No, most certainly not," his father said.

The electricity was off, and Evans started up his gasoline-powered generator and switched on his computer. "I didn't know how to use Facebook, but I know that everyone's on it," he said. Opening a group for locals, he posted pictures of all the dogs he'd found. Minutes later, he got a call from Karen Williamson and Stanley Jansson, the couple who had left their pug in their car when firefighters rushed them into the lake. Over the phone, Williamson was crying. *We thought for sure she was dead*, she told Evans.

The next morning Evans opened the door to a California Highway Patrol officer and a neighbor from across the lake. It was Scott Carlin. He knew of Evans's newfound role. "Are you the guy collecting neighborhood dogs?" the cop asked. There were two on the backseat of the cruiser.

Evans's personal menagerie was burgeoning. On the day of the fire, as he was still trying to save his own home, four mules emerged from the smoke. Evans patted them in greeting, and they came closer, putting their heads in the crook between his head and shoulder, as if for a hug. His dad gave them a peppermint candy and they stayed close to him until he led them to a neighbor's horse arena, where they would be safe. A dozen or so ducks settled on the Evanses' stretch of lakeshore, and twice a day one of them—whom they named Archibald—waddled up to the Evanses' back door for a bowl of feed and to wash himself in a bowl of water. A badly burned pig limped onto their property, and they put it with the donkeys. It soon died. They found another dead pig that they wished they could have helped.

Soon Evans started posting updates about which homes were

standing and which were not. He sought to stymie would-be looters by eliding identifying details—instead of an exterior picture he'd upload a close-up of the cuckoo clock inside. He received requests from exiled residents to see if their homes had made it. Even when the news was bad, people wrote back to him: *Thank you, at least now we know.* Their process of moving on could begin.

On the third day after the fire, as Evans was checking on a house, he saw a dog barking in the distance, across a deep, tree-lined gully in which a small creek flowed. As far as he could make out through the smoke, it was tethered next to a burned vehicle. From his truck he got a leash and a fistful of dog biscuits and made his way into the canyon. "A burned-up house and a volatile dog—stupid me," he said. He reached the other side and got about 50 feet from the dog, which seemed to weigh at least 70 pounds, when he realized it was not tied as tightly as he had thought.

He had a vision of bared teeth and the *tick-tick-tick* of the chain whipping up the ash as the animal went for him before he pirou-etted and high-tailed it back toward the gully. He seemed to have made it—was in fact jumping down over the crest of the slope—when it got its jaws into the back of his calf, just above his ankle, and buried its teeth. *I'm going to be mauled*, Evans thought. *How could I have been so stupid?*

The dog opened its jaws to clamp down again, and it was only Evans's momentum that kept him going and tumbling down the hillside. With a yammering heart he made it back to his truck and drove home, the blood streaming down his leg, puddling in his boot and spilling onto the floor. Limping into the bathroom just off the kitchen, he bent over, held onto the towel rack, and extended his leg behind him and over the sink so his father could get access to the wound. His father washed it, doused it again and again with hydrogen peroxide, slathered it with antibiotic ointment, and bandaged it tightly. Evans changed the dressing three or four times per day.

That was why, when he received a message from a local man about a dog the next day, he was wary. Arriving at the man's manufactured home, he soon found a black border collie with pale paws tied to a tree in the yard between a dog kennel and storage shed, both reduced to little more than white powder by the heat of the inferno. To protect herself from the fire she had dug a shallow hole in the dirt and tried to hunker down. By this point she had been outside without food or water for four days. Evans approached her slowly on his haunches. He shone a flashlight in his face so she had the clearest possible view of him.

A few feet away, he saw her tail wag. Afraid of provoking her, Evans didn't dare touch her, so he simply stood up, unclipped her 10-foot tether, and said, *Okay, come on girl.* Up close, he saw burns on her muzzle and ears. The fur along her back and stomach was scorched, and her tail was a solid matted or melted mass, like a club. At home he applied cold compresses, and soon whenever he opened the freezer door, she waited patiently next to it. He'd place them on her face or her back, and she lay perfectly still.

By this point he and his parents were living with eleven dogs. They chewed up shoes, slippers, bedspreads, pillows, and extension cords, and the puppies, which were not house-trained, irreparably ruined the carpet. But "you can't get mad at a puppy and you can't get mad at a dog," Evans said. Every day he did a round of homes where he knew there were animals, or where he might find them. He didn't want to cross the roadblock because he feared he would never be let back inside, but almost two weeks in, when he ran out of pet food, he drove down to the cordon at the intersection of Highway 70 and Concow Road. Nervously holding back amid the bustle of activity, he was trying to decide whom to approach when a woman turned to him and said, "You're Jeff Evans, aren't you?

We've been following you." His updates on Facebook had not gone unnoticed. The woman, who seemed to be managing the animal rescue crews, told him to take whatever he needed, and almost every day he came back for more.

They also asked for his help. They weren't permitted to break into houses, one animal rescue coordinator told him. And there was a rottweiler locked inside a home in Concow. Evans cringed. At the home, he managed to jimmy a screen off the window and open it. The dog walked past, growled at Jeff, and went into another room. Evans climbed into the house and ran to the front door, unlocked it, and slammed it closed behind himself. "You couldn't have got a needle in my butt with a jackhammer," he said. He gave animal control the all-clear.

He knew Lonnie Walker, the trucker who had gone out to work at 3:00 a.m. on the day of the fire, and his wife, Ellen, because he had done pest control for them. A rescue organization emailed him to ask if he could check on the couple's dogs. He didn't know that Ellen had died, and in the rubble of their home he propped up a metal sheet that he found and placed bowls of water and dog food underneath it. Every day, Evans came back to see if the water level in the bowl had fallen. He whistled for the dogs. But he never found them.

To keep his generator going and truck running for the month or so the power was off, Evans took gasoline from abandoned homes and cars. But he marked each can from the place he got it, so he could pay the resident back later. (It amounted to a little over $800.) His parents were imperturbable, and continued to play their two games of cribbage after lunch, as they had done for decades. But there were days when Evans would get home and he could feel the tension in his face, back, shoulders, and chest. Once in a while, tears welled up: "I don't even know what to say about that. There was no particular thought on my mind. The whole thing was a little overwhelming." He took comfort in the dogs. Some "would rush at me like I'm their owner, and I'd hug on them, kiss on them, and talk about a stress reliever," he said. Taking in the dogs "probably had to be one of the most medicinal things we'd ever done."

Over time, Evans learned how they came to be abandoned. And the animals' owners picked them up when they were able to reach Evans' home, with the exception of one. The owner of Bella, the dog who had been tied up outside and was badly burned, told him he thought Evans would be a better parent to the dog than he could be, so Evans kept her. Bella's wounds healed—she worried them with her teeth, pulling her fur out in clumps, until a thick, new coat grew back.

"She is the most wonderful animal I've ever owned," Evans said. "She sleeps at the foot of my bed, and in the middle of the night she goes down and checks on Mom and Dad, comes back, and lays on the floor next to my bed. Everywhere I go, she goes. I can't say enough about Bella."

The first meeting of the Paradise town council took place five days after the fire, at the Chico Council Chambers. Mayor Jones was living in a trailer and the other four members were with family, or in guesthouses and travel trailers. They joked they were a homeless council. That evening, they sat on a long, curved dais facing rows of folding chairs, a few empty.

Stepping up to a map of the Ridge on the wall, Paradise fire chief David Hawks gave an update on the blaze. Virtually the only fire activity in town was some smoldering stump holes, he said, and any vegetation in town that could burn had already done so, meaning there was no further threat.

Yet the fire's footprint continued to grow. At first, it had seemed like it might raid the valley. In its first twelve hours, it burned all the way downhill and jumped over Highway 99, a remarkable feat as highways are generally considered firebreaks. The fire approached the city of Chico, which issued evacuation orders for homes on its southeastern boundary, forcing some who had escaped the blaze in Paradise and sought refuge with friends and family to flee again. Firefighters cut firebreaks just outside city limits to protect the Stilson Canyon neighborhood, and later the seemingly endless black fields along the highway outside of Chico served as a reminder of just how close the Camp Fire had come. Even so, it continued to burn in the mountains, and firefighters streamed into the wilderness to engage it.

Testifying to the totality of the destruction in Paradise, Gill mentioned that the town's computer server had also been destroyed, so no one was receiving any email. "It's going to be a long, hard recovery, and some days it might get darker before it gets brighter," she told the room.

All stood for the Pledge of Allegiance, and bowed their heads for a moment of silence. A microphone picked up the sound of heavy sighs. About fifteen seconds in, the vice mayor, Greg Bolin, a large man in an orange polo shirt leaning on the table in front of him, offered a prayer with his eyes still closed.

Lord, we thank you for this opportunity to meet back together. We grieve so much for those that were lost. It really hurts. We pray for those that are in the midst of the battle out there with the fires. . . . Please put a hedge of protection around them, keep them safe. . . . Be with our people as they are displaced. . . . Give us the strength we need . . . to rebuild. . . . We love our town. We love You.

The councillors proceeded through business that had been put on the schedule before the fire—the mayor proclaimed it to be National Runaway Prevention Month in Paradise. "I apologize for the mundane in the midst of tragedy," she said. A handful of citizens took the floor. Paradise had recently passed a small sales tax for improvements, and one man said he felt he owed the town a proportion of everything he'd been forced to spend in Chico because of the evacuation. At the podium, in a touching gesture, he pulled the amount out of his pocket: $6.26.

Then a middle-aged man named Michael Orr stood at the microphone. He praised the performance of various city representatives and then addressed Mayor Jones. "You spent the last two months talking about how proud you were of your evacuation plan," he said. "Are you still as proud of it as you were then?" He called on the mayor to resign.

This was an intense moment for such a statement, when the mayor, homeless herself, was devoting herself to the recovery. "You can sit down," Jones told him icily. "I'm speaking to the room now."

"This is the same thing the media has asked me and I have said, *Yes, I am,*" she said. "It was chaos but it was sort of organized chaos. And I truly believe with all my heart without that plan, many, many more people would have died." It was a somewhat unusual step on the mayor's part—normally in council meetings, elected officials do not respond directly to public comment except to provide clarification.

"It's impossible for any town or city of any size to have the infrastructure to evacuate every person in their town or city all at the same time," she said. "You can't build infrastructure big enough to do that."

As the days passed after the fire and her father hadn't called, Skye Sedwick felt in her heart that he was gone. John Sedwick loved to talk on the phone, and urged his daughter to call and check in more. His niece, Simona MacAngus, wrote in a Facebook post that the family had heard from another relative of John's who believed that he was "out fighting fires with hearty younger men and women." MacAngus added that Sedwick had his cellphone with him, and always kept a charger in his car and his gas tank full for emergencies. "We do not understand why he hasn't had a fellow firefighter or someone send word to one of us," MacAngus said. It was a nice thought—Sedwick too busy to call because he was out fighting fires—but Skye didn't buy it.

The family reported him missing, and MacAngus talked to the Butte County Sheriff's Office almost daily. Several days after the fire began, they rang Skye. *We'd like to take a DNA swab,* the person on the phone told her. The first person she called was her mother, who lived on the other side of the country, in Pennsylvania. They got along well, but only saw each other every few years because of the distance.

If there is anything you need to say, she told her mother jokingly, *now is the time.*

Skye's parents did not marry and only dated briefly, and both Skye and her father had idly wondered whether, despite their physical resemblance, they were, in fact, biologically related. There was precedent for this in their family. Sedwick himself had long thought that the father he grew up with, Ralph, was not his birth father owing to his parents' open marriage. Ralph treated Sedwick no differently than his two older sisters, and Sedwick cared for him in his final days, but his hunch was largely confirmed when he received a call from the daughter of one of his mother's rumored beaus, who said: *I suspect you're my brother.*

Skye's mother didn't appreciate her quip.

"If I had to lie, why would I choose him?" she said. Skye's mother thought Sedwick was a good man, but despite their best efforts, they never had gotten along.

Skye drove up to Oroville early in the morning, and a lab worker swiped a cotton bud along the inside of her cheek. It was only after they took the sample that they told her they had found some remains on her father's property.

Nine days after the Camp Fire started, as at least 76 people were known to have died and about 1,200 were considered unaccounted for, Air Force One descended from the smoky skies at Beale Air Force Base, a 23,000-acre installation an hour's drive south of Paradise. Emerging from the plane in a camo baseball cap, black windbreaker, and khakis, President Donald Trump was met on the tarmac by California governor Jerry Brown and governor-elect Gavin Newsom, both Democrats and vocal critics of Trump. Brown had once said Trump would be remembered as a "liar, criminal, fool" for his scoffing attitude toward climate change. The president greeted Brown with a pat on the back and put his hand on Newsom's shoulder.

The governors declared in a joint statement that it was time "to pull together for the people of California," and flew north to Chico with the president and local congressman Doug LaMalfa on the Marine One helicopter.

Standing in a flattened mobile home park in Paradise, Trump returned to a subject he had previously brought up on Twitter. Two days after the fire started, before even offering condolences to the victims, the president blamed the state for poor "forest management" in a tweet. "Billions of dollars are given each year, with so many lives lost, all because of gross mismanagement of the forests. Remedy now, or no more Fed payments!" Trump was incorrectly attributing the fire to a failure by California to keep its woodlands in healthy condition by ensuring they didn't become overly dense with trees and brush. That is true of some of California's forests—but the Camp Fire made its big sprint from Pulga to Paradise in areas that were brushy rather than wooded.

Among desperate evacuees at the Walmart on that warm Saturday afternoon, there was little buzz about Trump's visit, though his motorcade was mere miles away, in the town they themselves were not allowed to return to. But even some supporters of the president said they were upset by his comments.

Kirk Ellsworth, whose adult children lost their homes in the fire, shook his head in disgust.

"My kids lost everything. I voted for him—and now? He can kiss my red ass," Ellsworth said, holding a bundle of stuffed tigers that he was passing out to kids in the crowded parking lot. "What he said was ridiculous. It hurts my heart. A lot of us voted for him and he [talks] down to us?"

Others said they hoped Trump's visit would draw attention to their plight. Ryan Belcher and his wife, Casey, were desperately trying to leave Walmart because in the days prior, someone had put up signs stating that the refuge would be closing and that evacuees should go to shelters or more permanent housing. The city, store, and Red Cross denied responsibility for the signs, but people began leaving nonetheless. The Belchers wanted another place to stay with their two children.

"We are not the ones to blame. We are not in charge of managing the forest," Ryan Belcher said, frustrated.

"I hope he sees how this community has come together," he added, trailing off as someone asked if he and Casey were fire victims, and then handed them gift cards.

"People are still here helping us. It's an amazing thing," Casey concluded, wiping away tears.

Up the road in Paradise, Trump addressed cameras against a backdrop of strewn rubble alongside local leaders—Sheriff Honea and Mayor Jones—and with Governor Brown and Lieutenant Governor Newsom. In Paradise he was not the brash, indelicate man who riled up crowds at rallies of faithful supporters. Instead the president took a softer approach and seemed more reserved. The praise he typically lavished upon himself, he offered to first responders: "We have incredible people doing the job, so we'll get that done better than anybody else could do it," he said. He even appeared ready to put aside political differences with California's leaders.

"To see what happened here, nobody would have ever thought this could have happened," the president said, hands on his hips as he stood in between Brown and Newsom.

"The federal government is behind you. We're all behind each other, I think we can truly say, Jerry," he said looking to Brown, who nodded. "We're all gonna work together and we'll do a real job."

Then he made an odd reference to a Nordic country. Such fires could be prevented, he said, if authorities had used practices similar to those he'd heard of in Finland.

"You gotta take care of the floors, you know, the floors of the forest, very important. You look at other countries where they do it differently and it's a whole different story," he said.

"They spend a lot of time on raking and cleaning and doing things, and they don't have any problem," he said Finnish president Sauli Niinistö had told him.

The idea that the solution to forest fires was a rake was ripe for takedowns on late-night talk shows and Twitter. And the Finnish leader denied he had said any such thing. "I mentioned [to] him that Finland is a land covered by forests and we also have a good monitoring system and network," Niinistö said.

Later that afternoon, while touring wildfire devastation in Southern California, Trump managed to compromise whatever goodwill his visit may have engendered.

"You don't see what's going on until you come here," the president said. "And what we saw at Pleasure—what a name."

Then for a second time he bungled the name of the town in his remarks. Jerry Brown had to correct him on what the town was actually called.

Richard Selden's team rented two RVs to work from and set up shop in the parking lot of the Sacramento County coroner's office, where the identification efforts were led by county coroner Kimberly Gin. Selden would end up spending the better part of three weeks there, along with three technical staff on the ground at any one time, working with four of ANDE's instruments.

In the early stages of the search-and-recovery effort, five to ten sets of remains arrived at the coroner's each day. Twenty-two of the bodies were identified by standard means, including, in two cases, thanks to serial numbers on medical hardware. Another eight were identified by ANDE when the team arrived, based on tissue samples that Gin had ready for Selden's team.

For the rest, a member of the ANDE team worked with the coroner's team to select the best sample for testing. Usually it was a minuscule piece of bone, about half a gram in weight. It was placed inside a plastic bag and macerated with a hammer to increase the surface area available for testing, and submerged for a few minutes in a solution that dissolved the calcium. A swab was dipped in the solution and placed inside the ANDE instrument, which required just over 100 minutes to analyze it. The results were uploaded to a database containing DNA samples from relatives of the missing, who had been encouraged to provide cheek swabs to police.

One set of remains was found commingled with those of two small animals, suggesting a person huddled with their pets in their final moments. On another occasion Selden produced a match, Googled the person, and came across a news story about their disappearance. But in the story, a relative of the victim was confident they were alive. It was devastating for Selden, who knew he was about to trigger one of the worst moments of the relative's life. The next day, he told Gin. "She said, *Don't do that.*" He recalled her saying that her own staff occasionally looked up victims, and she warned against it: "It can make you so sad that it makes you less effective. You don't realize it, but you're going through trauma right now."

His team members were tough, but they were not immune to the horror of their circumstances. One night a staffer was working alone in the lab, surrounded by remains, when the power flickered out. She called Selden and asked him to come over right away, and to work alongside her there in the future. After being on site for ten days, Selden wanted to fly back east to see his family for Thanksgiving. He arrived on Thanksgiving morning, but home felt foreign, and he thought he belonged in Sacramento because the work was incomplete. He took a red-eye back that night. In the end, ANDE was able to identify sixty of the victims and help ascertain the final death toll of eighty-five. Many had disabilities or were elderly. One of them was John Sedwick.

On November 15, the day after she drove up to Oroville to get swabbed, Skye got a call from the sheriff's office. They told her that the remains on the property were related to her. She could barely respond. For days she had been in agony, wondering where her father was and what had happened to him. Though Skye now knew he was gone, she also felt relief. He wasn't suffering or lost. Dad was at peace, she thought.

"We are collectively devastated because today I finally, after a week of uncertainty, [learned] that my father's remains were found at our house," she wrote on Facebook that night. "He absolutely died a hero trying to save our beloved home for me, my children and grandchildren to come. That house was his childhood home and he had so many memories there and on the Ridge. I hope you were blessed to hear him tell a story or two of his beloved Ridge."

There was another consolation. At least she now had proof that Sedwick was her dad. He always had been.

The list of missing persons was whittled down from over one thousand in the days after the fire until there was one name left on it.

Sara Martinez-Fabila was, on the day of the fire, fifty-one years old, a mother of five who moved to Paradise after five years in an alcoholism recovery program. "She wanted to become a lawyer, she was a very smart gal," said Lita Siebenthal, one of her seven siblings. But Martinez-Fabila relapsed, and when, after the fire destroyed the apartment complex they believed she was living in, the family did not hear from her, they thought that perhaps she had dropped off the radar as she sometimes did, or even that she was doing so well, thanks to all the aid they thought was pouring into the area, that she was off thriving somewhere on her own terms. Maybe, Siebenthal ventured, "she's got all the resources she needed at her fingertips, so she didn't need us." Their father was unwell, but once he was taken care of Siebenthal would figure it out. "I feel it in my heart that she's still alive, and I feel that as soon as we get my dad stabilized we're gonna—we gotta go look for her."

Several months later, her father passed away, and a year after the fire there was still no word from Martinez-Fabila.

What is there to be thankful for after you lose everything? Exactly two weeks after the fire, residents of Paradise were presented with this question.

At the Paradise Emergency Operations Center, the San Francisco volunteer Dave Sullivan had an idea based on his work in Santa Rosa after the 2017 fire there. He'd helped to organize a Halloween event for the displaced kids, and like others on the ground, it struck him that they needed to do something in Paradise for Thanksgiving. He blanched at the thought of serving people meals in the Walmart parking lot, so he wrote to a non-profit called World Central Kitchen, which he knew had provided meals after Hurricane Maria in Puerto Rico, to ask for help.

On Thanksgiving Day, the chefs José Andrés, Tyler Florence, and Guy Fieri cooked 7,000 pounds of turkey in a parking lot outside a Chico State auditorium. Food was served at a few other locations in the area—a campus dining hall and Sierra Nevada Brewing—so there was room for everyone.

At the auditorium, the buffet meals were served by a long line of uniformed firefighters and other first responders. They included Ken Lowe, the battalion chief who had herded dozens of residents into an intersection and probably saved their lives by doing so. "We're still here two weeks in, we haven't left the community," Lowe said. "To serve them a hot meal—that there is an honor for me." Lowe's mind still boggled at the destruction he witnessed. "I've never seen it in my career," he said.

Trestle tables covered in white tablecloths stretched the length of the hall, and people sat along them in bunches. At one was the wry Paradise high-school teacher, Virginia Partain, who saved her students' college admission essays. "I lost everything, so what else am I going to be doing?" she said, wearing the tie-dyed tur-tleneck she escaped in. She didn't have a home, or a classroom, but she'd met with students the previous day to work on their applications with them.

"I sat with my cats last night and I just held them," she said, "and I thought, it's a new chapter, a new normal, we just have to start a new life. I'm so grateful. Grateful I have my cats. Grateful I'm alive."

The mood in the hall was subdued and rueful. Survival had been assured for those in attendance, but it somehow continued to feel tenuous. As some of the firefighters themselves lined up to be served, spontaneous applause and whoops of appreciation broke out in the hall, and it went on and on.

When dawn broke the next morning, firefighters and evacuees looked up at the sky and thought, *thank god.* The storm that started the day before Thanksgiving continued to drench the Ridge. Over three days, an estimated seven inches of rain fell atop the charred houses, the blackened hillsides, and the still-burning hotspots scattered across the blaze's enormous footprint, which had grown to over 200 square miles. The lingering smoke was dispelled. By 7:00 a.m. on November 25, seventeen days after the Camp Fire had started, Cal Fire announced that it was at last 100 percent contained.

10

A Pile of Ashes

On December 5, under pewter, rainy skies, hundreds of residents motored through the parking lot of Paradise Evangelical Free Church on Pentz Road. Only drivers with a Paradise address were allowed past a checkpoint, and they were given hazmat suits, gloves, and foot coverings in case they planned to search the rubble.

Finally, the people of Paradise were being allowed to return. The next day, the police cordon around Paradise was removed, and anyone else who wanted to could enter.

The town they found was a charred and wintry wreck; 18,804 homes, businesses, and other structures were razed. Many of the trees were dead, stripped of leaves and frozen in the form they possessed on the morning of November 8. Even so, a strikingly large number were healthy, reinforcing the notion that the blaze that erased the town was an urban fire rather than a forest fire, and spread between structures rather than trees. Burned cars were everywhere, scorched hulks missing windows and tires. They had at least been pushed off the roadways; jagged scrape marks in the asphalt marked the path of the dozers that moved them.

Reminders of the town's normality were jarring. A number of

commercial buildings along Skyway survived, like the Walgreens and the Dutch Bros Coffee, though they had not reopened. There was a largely undamaged funeral home, the angel outside its door still an evanescent white. Of the few properties that did not succumb, many appeared dusty and neglected. Unseen owners or Good Samaritans had dragged fridges out to the curb and taped them closed to bottle up whatever foulness had putrefied inside after a month without power. At the trailer park that Trump had visited, the homes were all gone, outlasted by the garden gnomes in the front yard or a wheelbarrow planted with flowers. Deer quietly picked their way through the more immense silence.

Spray-painted markings on the cars and on the ground by buildings indicated if they had been checked for bodies, and if anyone had been recovered. At Holly Hills Mobile Estates, where at least three people died, the yellow tape and a leftover note from a search-and-rescue group still marked the spot where one victim had been found. Though his remains were gone, a charred set of keys and coins were left next to the burned wheelchair lift the man relied on to get in and out of his home. A couple who lived at the park returned and found that "there was clothing and still parts of flesh melted to rocks" inside one of the cordoned-off areas, said the husband, Albert Gurule.

In the coming months, more than a thousand volunteers from an evangelical Christian relief organization worked in Paradise, helping hundreds of homeowners sift through the ashes and praying with them.

"This was somebody's hopes, their dreams, and it's gone," Samaritan's Purse president Franklin Graham said on a visit to the area.

People searched for heirlooms, antiques, jewelry—but the tragedy had laid bare every aspect of their lives, including, inevitably, belongings that were considerably more intimate. One January afternoon, the volunteers, wearing plastic coveralls over bright orange shirts, helped a young college student sift through the rubble of her home. As they searched, she thought, *I hope my dildos didn't survive.* They worked for about two hours, and when they finished she breathed a sigh of relief. They asked her to pray with them, and she wryly thought to herself: *I'm still an angel in their eyes.*

For survivors, the grief was too much to bear. Yet it had to be borne. Children wouldn't sleep. A woman whose father died could not return to work.

In the days after the fire, Iris Natividad had been staying in hotels in the towns of Lincoln and Yuba City, but now she was sleeping on an airbed on the floor of an unfurnished house that was not her home. Waking one night, stumbling to the bathroom and back, she felt utterly discombobulated.

Where the fuck am I? she thought.

Natividad was in Bangor, the community she had lived in for ten years with Andrew Downer before moving to Paradise. She was staying at her former neighbor's place—her old house was in fact right across the street. Just six months before the fire, the neighbor had offered it to the couple to rent, and they seriously considered it, thinking it was so large and airy, at 4,000 square feet, that it would be the perfect place to display their antiques. They were deterred by the same factors that prompted them to leave Bangor in the first place: it was too isolated for someone in precarious health.

It was hard for Natividad to think or plan for the fog in her brain. A creature of routine, she didn't know what to do with herself now there weren't the same therapists' and doctors' appointments to ferry Downer to, the same restaurants to eat at. "I actually went to Paradise a lot, to the property," she said. "I think I wanted to be close to Andrew." Other times, "I would drive and I didn't know where I was going."

At the grocery store, she saw someone with the same brown Coach bag that she had. Except she didn't, she reminded herself. Not anymore. One of her friends asked her where she was with the stages of grief. Had she experienced the anger yet? As it happened, Natividad had just gotten mad at her washer/dryer. She already owned one—*used* to own one, she corrected herself—so why did she have to pay $1,000 for another? To receive an insurance payout, she, like many others, was asked to make an inventory of everything she had lost. So she was trying to go from room to room in her head, looking at old photos, scanning eBay to jog her memory. The "hardest thing I've ever done in my life is to make a list of all you have lost in your life," she said. "The trauma is unspeakable."

People couldn't truly grasp what she was going through. "Everyone says, *are you excited to buy new furniture, a blank slate?*" Natividad said. "And I'm like, *no, I'm not happy about it.* We were antique dealers, we had thousands of items that we cherished. Everything had a memory."

In addition to attending grief support groups, Natividad began seeing a therapist in January. It was actually Downer's therapist, who had made such a difference in his life and helped him see that unhappiness was not the only option. She had lost her home, too. In their first session, they talked, Natividad said, only of Downer. In later sessions, they turned to the town, and the grief of losing it. After that, they turned to Natividad herself.

It seemed to Natividad that she was not experiencing grief the way she should be, the way her friends were. Work was her coping mechanism—she was dealing with other people's problems, not her own—and she never stopped going. She felt she could function. But then she realized that she was not, in fact, able to look at Downer's picture. And she would be triggered by memories, objects, things she saw by the road, and she would begin to sob.

Her way of dealing with these feelings and reactions was, as best she could, to shut them down. "I squint my eyes and what I do is, I shake my head like, *No, we're not doing this.* I shake my head and squeeze my eyes and squint them and bear down and say, *No. And that's how I stop the PTSD.* I stop that emotion. I don't let it affect me." Natividad knew that she was avoiding something.

Gradually people began to return to Paradise on a more permanent basis, creating little hubs of activity in the emptiness.

Lyons Express Lube and Oil, just off Skyway, was among the first businesses to reopen. Its seventy-two-year-old proprietor, Paul Lyons, had run the business for thirty years, and he'd fought the flames during the fire for about eighteen hours, as gas tanks from the heating and air conditioning store next door exploded and flew above him. Nearby, his home had burned down.

Turning on the power was as easy as flipping breakers. He brought in a five-gallon tank for drinking and washing. His wife Linda posted about their return on Facebook. The couple trusted in their clients' faithfulness. "We will follow our doctor, our hairdresser, and our mechanic anywhere because we trust them with our bodies and our car," Linda said.

On January 2, Lyons's first customer of Paradise's new era was a family in a station wagon who showed up at 8:00 a.m. for an oil change, and who demolished the celebratory donuts, muffins, and coffee he had set out on the jukebox. They looked to have all their belongings in their car and hadn't eaten since the morning before. Business that day was sparse, but over the ensuing weeks Paradisians drove dozens of miles from surrounding towns just to give him their custom. Surprisingly, some days it was as busy or better than it had been before. "I guess I've always been the kid with the shovel," Lyons said, "and I know there's a pony underneath that pile of horse manure."

Paradise Bikes cracked its doors around the same time. A pizza truck—called, unfortunately, Inferno Wood Fired Pizza—parked on a burnt lot and served cleanup crews and whoever else was passing through town.

One weekend, with dead pine trunks against a cobalt sky and a delicious warmth seeping back into the air, Iris Natividad drove back into Paradise. She passed the eighty-six white crosses that Greg Zanis, the founder of a Chicago-based organization with a sad niche, Zanis Crosses for Losses, had built and erected at a prominent corner on Skyway back when the number of fatalities fluctuated slightly based on new evidence. A flashing billboard advertised help with rebuilding. At the north end of downtown, she arrived at a little chalet-style commercial building, with wooden railings and a lawn out front. Most other structures around it were gone. A large sign proclaimed its name: Treasures from Paradise.

The store was owned by a couple who lived in Chico, Rick and Barbara Manson, who jointly operated the cash register next to the door, recording sales by hand. It was run like a cooperative, with several sellers who stocked the shelves with antiques, furniture, and thrift store knickknacks, ranging from vinyl Madonna and Eagles records to decades-old issues of local-history newsletters. Its previous location had burned down, and this was the new store's soft opening, with coffee and cookies for visitors. The actual opening was planned for a day later, and a local TV news crew was expected to show up.

Natividad had a whole small room to herself, in which the star attractions were a cabinet full of uranium glassware and painted wooden signs she'd had made that read "Paradise Strong," the slogan that had become the rallying cry for the recovery effort. Natividad hadn't really known if she was going to set up shop until she did. All her friends had moved—to Oklahoma, Idaho,

Nevada. Natividad had been thinking about moving as well, for instance to the Oregon coast. There was nothing left for her in Paradise, it seemed. She had a rough timeline in mind: "My therapist always says, *Give it a year.*"

A man walked into the store clutching a cup from Dutch Bros Coffee, the regional chain whose Paradise outpost reopened that same day. "It's so good to see you up here, so good," he told the owners.

"Well, we're doing what we love," replied Barbara Manson. "We're putting one foot in front of the other and keeping going."

It was a gamble for the couple. Their potential customer base in Paradise had shrunk by over 90 percent. But more homes had survived in Magalia, north of town, and the store was on the thoroughfare that led to it. "People still have to replenish their Pyrex, dishes, the stuff from before," said Barbara. "That's what we do, what we have to help them do."

That day Natividad sold a number of the "Paradise Strong" signs. Half a mile away from where she chatted to customers in the store was the home she'd shared with Downer.

Taking a break from the shop at lunchtime, Natividad stood in her front yard, dressed in cropped jeans, a simple black blouse, and a necklace with a butterfly pendant, and showed the wire-mesh sifters that she and some friends had used to hunt through the rubble for anything they could salvage. They'd unearthed some tankards decorated with Norman Rockwell images, and they'd been able to recover a lot of the marbles that Downer kept stacked in the closet. "We have buckets of them that survived," she said. "What do you do with them? I'm not sure."

Mostly everything else, however, seemed to have been reduced to junk. "At this point it feels like it's time just to leave it," she said.

At the top of the ramp that led up to the home, next to where the front door and the gumball machine of marbles had been, she pointed out a stain in the concrete a few square feet in size. The black whorls looked like inkblots that had bloomed on a damp piece of paper. This was where Downer had died and where the flames had consumed his body. A broken piece of his wheelchair still lay on the ground.

She stood on that particular part of the path and took in the scene around her. At times, the house was a gut-wrenching place to be. Then again, "I feel this is where his spirit is," she said. "I feel warm and fuzzy—actually I feel good when I come here. I feel his spirit and the dog's spirit. I feel close to them."

Scattered everywhere, catching the sun from their bed of ashes, were fragments of ivory, turquoise, and uranium-green glass. All that Natividad and Downer had collected—and even particles of Downer himself—had become jumbled and fused together in a rare amalgam. The history concentrated in this one spot, of Natividad and Downer, of the antiques with their own life stories, was dense and tangled.

Natividad had a friend who was a medium, and although Natividad didn't ordinarily believe in such things, the message given to her by the friend was a comfort.

"She says Andrew took everything he loved with him."

Though many wanted to return to Paradise, the obstacles seemed almost insurmountable.

In the weeks after the fire, FEMA, the federal agency tasked with coordinating the response to US disasters that overwhelm the resources of local and state governments, announced plans to provide as many as 2,000 temporary homes to those displaced by the fire. The agency set up shop at the former Sears department store in the Chico Mall. Some Paradisians waited there for hours, in a queue that started outside and snaked through the massive facility, to speak with agency representatives and request aid in the form of cash grants and temporary rental-housing assistance, as well as reimbursement for the cost of hotels and motels. These had no vacancies for miles around and for months after the fire their elevators and lobbies were full of people whose faces suggested the same wearied determination, the same lost capacity for surprise, as they sought to navigate insurance and federal bureaucracy while crammed with their families into shoebox rooms with no place to go back to.

Crews began clearing the remnants of nearly 14,000 destroyed homes in January, scraping them down to their foundations, a process they warned could take at least a year. Uncleared lots were unsafe to live on because the debris was considered hazardous, filled with toxins like paint, burned batteries, and asbestos.

The town passed an ordinance in February making it illegal for residents to live on property that hadn't been cleared of burned debris. FEMA had warned Paradise that if it allowed people to remain, it could lose the $1.7 billion allocated toward cleanup costs. So more than a hundred residents who had returned to the town in RVs and trailers parked on their lots had to relocate.

This news was another blow to people who'd already lost everything. Anastasia and Daniel Skinner purchased an RV for themselves and their four kids with the understanding that they would be able to stay on their property. The announcement that this was not so sent Anastasia into a panic. "I didn't think it was possible for them to kick us off our own land," she said. "If there was any other place to go, we would be there."

In early spring, the Paradise Irrigation District discovered that benzene, a cancer-causing chemical in crude oil and gasoline used to make plastics and synthetic fibers, contaminated almost one-third of the five hundred water samples it collected around town. Officials believed this occurred when the Camp Fire sent toxic gases from burning homes into the water pipes, and melted plastic pipes and meters. By the summer, the water district was still warning people not to drink the water or use it for cooking, brushing teeth, making ice, or preparing food. Residents were told to limit shower time, use cold water to wash clothes, and not to give the water to pets and livestock. "Because of the nature of the potential water contamination, boiling or disinfecting your water will not make it safe to drink," officials said.

The competition for living space was fierce. Two hundred homes sold in Chico within three weeks of the fire, some to people who knew their houses were gone, others to people who had no idea but weren't taking any chances. In Chico, many rentals were already taken by some of the 17,000 students from the university, and Butte County, like most of California, had a housing shortage. The rental vacancy rate was between 1.5 and 2.5 percent before the fire; after, it dropped down to near zero. Authorities at the county housing authority estimated Butte County could absorb between 800 and 1,000 people. While the full extent of the destruction was not yet known, Ed Mayer, the county housing director, knew that many of the displaced would be unable to find housing locally. "I don't even know if these households can be absorbed in California," he told reporters.

It was difficult to precisely track the whereabouts of Ridge residents, but of the roughly 50,000 who evacuated, 23,000 applied for financial or housing help from FEMA. In these cases, FEMA documented their ultimate address, and learned that some 16,500 remained in the county, while the rest migrated throughout California and across the United States. Oregon took the most of any state, with nearly 250 former Butte County residents moving there, while others settled as far away as Vermont, West Virginia, Puerto Rico, and Micronesia.

David Micalizio was one of almost twenty people to settle in

Utah. He had moved to Paradise in 1995 to care for his ailing grandmother, and when she died, he decided to stay. An arborist, he had no shortage of work, and had only just retired when the fire hit. It destroyed the 900-square-foot home built by his grandfather in 1956, and in the ruins he and his brother found that even his loose change had melted. Among the few items they recovered was a brand-new cast-iron skillet that Micalizio's brother had gotten him a few months before. It had a hole in the middle. He didn't have insurance and wasn't going to rebuild.

His daughters, who both lived in Utah, came for Thanksgiving and invited him back. As he sat on the train pulling out of Sacramento, it felt as if he was headed to another country—he had never lived outside California and felt most at home in Paradise among the trees. *Here's the next journey*, he thought.

Rachelle and Chris Sanders had decided to stay in Chico with baby Lincoln, but it proved to be a calamitous year. When Rachelle and her family were moving into a new apartment, she noticed that her husband seemed short of breath and unable to keep up with his father-in-law. Sanders urged him to go to a doctor, and he was diagnosed with severe anemia. Soon the couple learned it was worse: he had leukemia. He had a bone-marrow replacement, but within three months he died. In less than a year, Rachelle Sanders had lost her town, her home, and her husband.

The parents of Kimber Wehr, the woman who couldn't get out and whose voicemails described an "apocalypse," had left a last, dreadful voicemail for them, moved near Santa Cruz, where they'd bought a condo decades before with the hope that Kimber and their other daughter might live there. "What's good is that we already knew a great many neighbors," Pat LeBlanc said. "We were given a warm and compassionate reception."

Even though they missed their routine, it was a blessing to be somewhere that still felt like home. Visits to Paradise to wrap up their old lives weighed heavily. "I will avoid going there for years if I'm so lucky as to have that opportunity," she said.

Lonnie Walker, the man who ran the roadblock to find out what happened to his wife, Ellen, bought a home in snowy Plumas County, 90 miles to the north. It was closer to his employer, and it was a relief to escape the memories. But later he moved to the town of Oroville, much closer to Paradise, to be near the rest of his family. "I was trying to run away, but I realized that I can't, and I enjoy frequenting and seeing the places that my wife and I frequented," he said.

The restaurants, the pharmacies, where she got her hair done— "it's just nice to do business where we'd done business all those years. I don't know if that's good or bad, but it's good right now for me to try to work through this thing." He resumed trucking— the very job he was about to give up, at Ellen's request, when the fire struck—but had to stop again after suffering a stroke. He got remarried, to a woman he met at church.

Ten months after the fire, he received painful news he was unable to make sense of. The Butte County sheriff announced that a set of remains had been found commingled with Ellen's at their home on Schwyhart Lane. It was a "larger older adult male" who had "dental work with crowns," according to anthropologists. There was no information on who he might be, and no extant missing-person report. Lonnie wondered whether it was someone from a pot grow near their home who had tried to save his wife's life. But in truth, "I know that I just don't know," he said.

Bill Goggia, the man who was rescued from the roadway by two men in a pickup and who lost the sight in one eye, was staying with his eighty-seven-year-old aunt, Dorothy, who lived in one of the five homes that had not burned in a neighborhood of perhaps seventy. It was right next to Magalia Cemetery. The surroundings may have been a mess, but inside it was as if nothing had changed.

There was the heavy beige carpet, pine walls, a Bible on a table next to a glass bowl of candy. "My late husband and I bought this place in 1957," Dorothy said. Now she had what she called "survivorship guilt." "It's hard. I don't even have to look outside—we don't have traffic."

Goggia walked with a limp, one eye clouded. His aunt's place wasn't far from his own burned home, and he went back there all the time, rummaging through the debris and looking for the metal urn that held his father's ashes. "I'm a pretty positive person, but even when I'm going through town . . ." he said, trailing off. "I think it's gonna be a long time before you can drive through here and not be depressed about the scenario. It's like a bomb went off."

The dislocation could be deadly, especially for elderly evacuees. They died of cancer, strokes, and heart attacks, but their families all believed that the stress of the fire, and of seeing the Ridge destroyed, played a part. No one knew exactly how many, as no agency tracked that information.

After finding his mother, Barbara Allen, at the Walmart two days after the fire first tore through the Ridge, thanks to the townspeople at the community meeting, Robert Edwards spent more time with her than he had in decades. Edwards had moved from the Ridge in 1991, and in the last decade they had spoken over the phone once a week and saw each other once a year. Allen had never been to her son's home. The feisty, self-reliant woman devoted most of her time to volunteer work in the community, Edwards said, working at the senior center and playing cribbage. Now she moved to his place in Seattle, where he worked as a senior engineer for a tech company. Each night she'd watch *Wheel of Fortune* and *Jeopardy!*

From the beginning, he could tell something wasn't right with her. She seemed to be in pain constantly and was unusually forgetful. A visit to the doctor revealed she had cancer that had spread to her liver and kidneys.

They had just three months together from the time he found her until she died on Valentine's Day. And it was only when Edwards announced his mother's death on Facebook that he grasped her true impact in Paradise. Messages poured in, paying tribute to her. "Everybody knows her. Everybody said she did everything possible for them," Edwards said. "It's like she lived this completely different life in Paradise by herself that didn't include her kids."

They buried her next to her older sister, near a Korean dogwood at the Paradise cemetery, on an unusually warm April morning a day before what would have been her seventy-eighth birthday.

A collection of people who loved Barbara Allen most stood in a circle, wiping sweat from their brows or fanning themselves, and trying to find a bit of shade under small trees as they waited for the preacher. Enough time had passed since Allen's death that the group smiled and laughed as they shared memories of her or noted how much Edwards resembled his mother.

When the preacher finally arrived, they gathered in front of her gravestone, etched with a bird and the inscription: "Beloved sister, wife and mom." A bright pink and purple bouquet sat on top of it. Allen had paid for her plot and funeral expenses fifteen years earlier. Some elderly people struggled to stand in the heat as the service progressed, but they made no move to sit—they were determined to honor her.

If Paradise came back, Edwards said, he would help rebuild the senior center. In fact, he'd put her name on it.

"The town of Paradise dying, it broke her heart," Edwards said. "I don't think she had the will to live after that. There was nothing else for her. She loved it that much, she really did. This is where she wanted to be."

Most of Paradise's institutions shrank, dissolved, or regrouped elsewhere. More than five thousand kids from Paradise, Concow, and Magalia were out of school for nearly four weeks immediately after the fire, along with tens of thousands of others in Butte County. Magalia's high school was destroyed, and in Paradise an elementary school and a small joint elementary and secondary school were incinerated. Paradise High was left standing, but owing to smoke and fire damage, and the overall state of the town, its students had to relocate for the remainder of the school year. "I promise you we will graduate a Paradise High School class of 2019 this year," the district superintendent reassured anxious parents in the days following the fire.

In December, elementary school students reunited with their teachers and classmates at schools lending them space in Durham and Oroville. Middle-school and high-school students completed most of their coursework online and met with teachers and classmates at the Chico mall.

About five hundred of Paradise High's nine hundred students relocated in January to a temporary campus in northern Chico that students called "the fortress" because of its location on Fortress Street. It was so important to some of the students to be there that they commuted from more than an hour away; others separated from their families and couch-surfed with friends. The other four hundred students left the district or opted to take Internet courses—some were too traumatized to return.

Arissa Harvey, the student who had escaped Paradise High on the morning of the fire, got a ride to the Fortress every day from Oroville. Her family's home in Paradise had burned down, and she and her two sisters had ridden out the fire in their metal barn, huddled under a blanket and trying to suck in the air from an oxygen canister they opened because the smoke was so oppressive. Her first lesson on her first day back was Advanced Placement history with a kind teacher. "When we walked into class he was like, *I know a lot of things have happened, I know you guys are everywhere and there are not a lot of resources right now. But we have an AP test in a couple months and we're gonna pass that AP test.*"

She passed the test, but otherwise felt she was floundering. She prided herself on being a mostly straight-A student, but at the Fortress it was hard to focus because the classrooms were divided by movable walls not much taller than an adult, so the Spanish conversation class was painfully audible to the class trying to take a silent math test next door. There was nowhere to do homework. To Harvey, it seemed that real life, where doing well in school could have specific, positive consequences, had been interrupted, and there was nothing to look forward to. The teachers gave students a pass for behavior that normally would have earned censure, like using mobile phones in class, and prompted them to talk about what they experienced in the fire. To Harvey that felt awkward and private.

As the superintendent had promised, every senior would end up graduating. But there was an unexpected casualty.

The home of Paradise High principal Loren Lighthall had burned down, and he was living in a small three-bedroom apartment in a rough part of Chico, near the university, with his wife and five of his seven kids. The complex was close enough to the railroad tracks that they woke every night at 2:00 a.m. when a train passed. Drunk neighbors occasionally banged on the door late into the night, mistaking the apartment for their own.

Unable to find an affordable home for his large family, Lighthall announced in April that he had accepted a job at a high school near Modesto, more than three hours south. The news was a blow to students, and some offered to find Lighthall housing. But he was ready to leave, not least because everywhere he went it seemed the fire was the only topic of conversation. He had to remove himself from some of the survivor Facebook pages.

"It's kind of like having your mother's gravestone in your living room," he said. In the first three months at his new job, hardly anyone asked about the fire, and he could go weeks without having to talk about it.

Paradisians didn't just lose their homes, they lost their liveli-hoods. One of the most ruinous blows to the cohesion of the community was the pullout of the town's biggest employer, the Feather River hospital. It was closed immediately after the fire, and the company that owned the hospital, Adventist Health, later announced that it wouldn't reopen the hundred-bed facility, lay-ing off more than 1,300 staff. It paid workers from the time of the fire through January.

Chelsea West, the nurse who was among the group rescued by dozer operator Joe Kennedy, was splitting her time working at an oral surgeon's office as well as a medical spa, and teaching yoga in Chico, with no plans to return to a hospital setting. Though she loved the work, she didn't want to be taken back to that day. Three weeks after the fire, with a boot still on her broken foot, West headed for Washington, D.C., with climate-change activists and a local congressional candidate. They planned to speak with environmentalists and Democratic politicians.

At the Dirksen Senate Office Building, she met with Bernie Sanders. "We were wearing T-shirts the days before the fire," she told him. "And it should be raining. Things are changing quickly. And every single year it's worse and worse. We want to help pre-vent this from happening to other people and other communities because it will."

The grizzled senator, wearing a navy blazer and speckled blue tie, looked astonished by her story. "Whoa," he said, eyes wide, his hand over his face.

After Allyn Pierce, the nurse who thought he would burn up in his vehicle, shared pictures of his toasted truck on social media, Toyota gave him a new one and invited him to a factory in San Antonio, Texas. He and his young daughter were asked to come on stage at a Kelly Clarkson concert. California's new governor, Gavin Newsom, who succeeded Jerry Brown in January 2019, invited Pierce to his state of the state address at the capitol in Sacramento and recounted his exploits to lawmakers.

Pierce, wearing a flannel shirt, looked uncomfortable when those in the chamber began to applaud him. He appreciated the recognition, but the word "hero," which appeared in countless headlines, made him uneasy. He believed he was just doing his job.

Paradise and the hospital were gone, as were his friends, and when the extent of the damage became clear in the weeks after the fire, Pierce knew he wouldn't return. His family moved to Chico, a place he'd always found livelier than Paradise, and he started working at Enloe Medical Center. It gave his family necessary stability. He missed Paradise and his colleagues but grew frustrated when people suggested that he should rebuild out of a sense of allegiance.

"I have loyalty to the people. The town doesn't exist right now," he said. "I don't have loyalty to a pile of ashes."

During the two weeks the fire burned, not a single police officer missed a day of work, according to the police chief. But in the months after, some moved out of state, and others, concerned about job security in a town with a diminished tax base, took positions at different departments. A year after the Camp Fire, the department had lost ten of twenty-one officers and all but one dispatcher, and had to contract some dispatch services out to the county.

Kassidy Honea, the officer who got a hug from her father, Butte County Sheriff Kory Honea, during the fire, decided to move to the sheriff's office in Oroville. She had initially hesitated—she didn't want her peers to think she had gotten the job because of her dad—but as friends and mentors left Paradise, Honea changed her mind. Paradise had grown too quiet for her.

Generally it was the older officers, further along in their careers, who intended to stay, such as Rob Nichols, who had helped establish the temporary refuge at the Optimo Lounge. After the fire, Nichols, his wife, and two kids spent nearly a year living in an RV amid rolling, golden hills 10 miles down the road from Paradise. They got cabin fever from time to time, the four of them trying to coexist in a space a quarter the size of their old home. Some weekends they'd leave the RV behind and head to the coast, where they'd get a motel room that felt spacious by comparison. Nichols thought he might return to Paradise when the cost of rebuilding decreased.

In the early days, working in his incinerated town could be depressing, and he felt flashes of anger over what had happened to Paradise. Nichols also felt like he had grown. He had spent nearly twenty years being exposed to people at their worst, but in the fire he saw acts of generosity: neighbors sheltering neighbors, strangers who donated millions. "Everybody took care of everybody," he said. "It kind of restores your faith in humanity."

Law enforcement in these circumstances had its own tenor. As officer David Akin turned left on Valley Ridge Drive on a Saturday afternoon in February, the rubble gave way to manicured grass and long driveways leading to immaculate homes. Two houses, one blackened lot, two houses, and then another blackened lot. Akin, who had lost the house he shared with his wife and young kids, spotted a couple eagerly waving at him.

He had been with the department for four years when the fire hit, and was still in the early part of his career when many officers are eager for the adrenaline rush of "chasing bad guys," he said. This couple just wanted to make his acquaintance. "We live just over there," they told him, pointing. "The one with the big flag out front if you ever want to come say hi."

Other officers encountered people who seemed to think that

because the town was mostly gone, the speed limit no longer applied. They arrested drivers for DUIs and responded to calls about suspicious people on burnt-out lots. Even so, Paradise had suffered so much, some officers said, that it required them to be more compassionate.

"It's a whole different kind of policing," Officer Matthew Gates said. "Unless you do something really outrageous and blatant, I have a hard time giving someone a ticket whose house was lost, or whose house was standing but their community is gone, or people who are trying to help the town rebuild."

It took about a month, but finally Skye received FEMA money for hotels and was placed on a list for temporary housing.

In December, her father's funeral was held at his church in Oroville where thirty-five people spoke, some of whom Skye had never met. They shared stories of Sedwick's kindness and generosity. One man said that when he told Sedwick he hoped to learn the guitar, but didn't have an instrument himself, Sedwick gave him one. "I didn't like John when I first met him, but I grew to like him a lot," another man said. Skye valued his honesty—she knew her father could be ornery. She didn't appreciate a preacher who had said, in private, that she should have told her dad to leave. "He wouldn't have gone," she told him.

FEMA delivered Skye's 300-square-foot camper trailer in January. She moved in with the dog her dad had given her, Jude, and her few belongings: clothes, including the gray skinny jeans she'd worn on November 8 and now called her fire jeans, her gray Chucks, and some sweaters. Her cousin Simona MacAngus gave her dishes and rugs.

In March, Skye shuffled around her trailer, from the table covered with paints and canvases to the kitchen a few feet away. With Jude trailing behind her, she picked up a vase of flowers and threw them away.

"Everything dies in here," Skye said, shaking her head. "I can't figure out why."

The small trailer now had a few items she'd picked up at swap meets as well as photos of her father and stories he wrote for a local-history journal, *Tales of the Paradise Ridge*. She wasn't allowed to hang anything in the trailer, so instead she placed miniatures, tea-cups, and figurines on the table and counters. Peacock-patterned scarves hung over the windows.

Her new home was a far cry from the cabin she and Sedwick shared overlooking the canyon. Orchard RV Park was located just off Interstate 5, the busy highway that spans the state, in the small agricultural town of Orland. Here there were no trees to be tended to, antiques to be dusted, or meals to be cooked for Sedwick, just forty other trailers filled with people like Skye who had lost their homes and their town and were trying to figure out how to pick up the pieces. Sometimes it was the little things she missed. She couldn't find any underwear that was quite as com-fortable as the underwear she had before.

This was the longest time in her fifty years Skye had ever lived by herself. It was lonely, but if she forgot to put the toothpaste cap back on, or didn't feel like getting dressed, no one was there to notice. Skye tended to keep to herself in this park of fire refugees. Work-ing as a counselor in a high-security juvenile detention facility, she had seen the conflict that could occur among groups of traumatized people. "My experience has taught me that when you have a lot of people with PTSD, it's not a real healing place," she said

Skye spent most days in the cool trailer with Jude, trying to get her Magalia property cleared and figure out her new life. When she wasn't attempting to cross things off the to-do list on her legal pad, she painted and looked for cleaning work online. She had been taught by her mother, who worked in a veterans' hospital, and was good at it—detail-oriented and personable—but it was hard on her body.

Some days Skye didn't make it out of bed, let alone out of the trailer. She was lonely. She picked up a stutter she never had before, and fell into a deep depression. Skye thought about driving home to Magalia, to the property, but she would add up the cost of gas in her head and realize she didn't have the money to make it there and back. Occasionally, she'd go to the casino 11 miles up the highway. It helped relieve the solitude. And some weeks, the money she won there was her only regular income. "It's terrible," she said. "You don't want the people who help you to know you've done that. There's shame involved with that."

She thought of her father every day.

Once a month a FEMA worker stopped by to inspect the trailer. They looked around and made sure things were in order—one once took a photo of the inside of her fridge. But mostly they came to discuss what Skye would do next. It felt to Skye like a polite way of saying, *When the fuck are you leaving?*

She didn't know what to tell them because she didn't have any answers. Some days she thought, *I'll rebuild.* But she also wondered if she could ever afford it or if it was safe, considering the risk of contaminants and yet more fires.

When she had dark days—friends didn't reach out, her kids didn't call, and it felt like no one in the world was thinking about her—at least there was Jude. He stayed with her while she painted, or read, or when she would simply sit. Most people don't just *sit* anymore, without any other purpose in mind, Skye thought. She would settle in the chair in her living room and pet Jude, his big brown eyes looking up at her.

In the midst of her grief, for her father and her home, Skye was working on her lawsuit against PG&E. She had picked a lawyer— someone local who had also lost his house. Like a soaring number of others, she blamed PG&E for upending her life. It needed to be held to account.

11

The Perfect Fire

The Camp Fire had given Paradise a dubious sort of global distinction. It was the world's most expensive natural disaster of 2018, causing $16.5 billion in losses, $12.5 billion of which was insured, according to Munich Re, a German insurance company that tracks major catastrophes. It was costlier than the Sulawesi earthquake and tsunami, which left at least 4,300 dead, and Hurricane Michael, a category five storm that led to the deaths of at least fifty people in Florida.

The scale of destruction and unfathomable price tag—more than a thousand times larger than the town's annual budget of $13 million—prompted a pair of questions that echoed from the FEMA trailers in Orland to the Paradise town hall and the shores of the Concow reservoir. Who was to blame, and who would pay?

Word traveled fast that PG&E had reported the outage on its Caribou-Palermo transmission line near the Camp Fire origin point. Though residents at one of the early community meetings had loudly insisted a PG&E executive be treated with respect, they were suspicious. They had not forgotten the deaths of eight people in the San Bruno pipeline explosion and 2017's deadly

North Bay fires. Among Paradisians, it didn't take long for PG&E to become a byword for negligence and corporate greed.

At the official memorial for the victims, which took place three months after the fire at the Paradise Performing Arts Center, a newer building that had survived, there was little mention of who might be to blame for the disaster. The darkened, 760-seat auditorium was full, and attendees stood crowded near the doors at the back. On stage, the decoration was minimal, little more than the outlines of a few trees in Christmas lights.

There were twenty speakers representing a kaleidoscope of credos: leaders of various Christian denominations, as well as a rabbi and a Scientologist. Prayers from the Haudenosaunee, Rastafarian, and Baha'i traditions were intermingled with a reading from the Koran and a gospel choir performance. Seven women wearing blazers and long skirts who belonged to an international Buddhist nonprofit presented an elegant dance. A local couple sang a song that included a line warning looters seeking spoils in Paradise that they would find "only bullets," which drew an appreciative laugh from the audience.

Pastor Krystalynn Martin, a Paradise native, read a poem of hers that had circulated widely on the Internet. "Please excuse the smoke," it began. "It's just the dreams and hopes of 27 thousand yesterdays." A speaker from the Pagan community sought to point out the good in fire—"Fire transforms, fire purifies, fire is like the light of the soul, always changing, growing"—which, despite being accurate in ecological terms, made for an uncomfortable moment.

And during a piano solo, the names and images of the victims were projected onto a screen for a few seconds each. Afterward, at least one attorney was milling around, and indeed all along Skyway, among billboards thanking first responders or proclaiming "#RidgeStrong," signs had appeared trumpeting the credentials of this or that lawyer. Even Erin Brockovich herself had been in town to stump for a group of attorneys, and was trying to sign up clients, as she had decades before in Hinkley.

What appeared to be the first lawsuit against PG&E had been filed just five days after the fire began, even before investigators had officially determined the cause. It was brought on behalf of several families and individuals, comprising approximately twenty-five people. There were few personal details in the filing, but it stated: "Plaintiffs have suffered and/or continue to suffer great mental pain and suffering, including worry, emotional distress, humiliation, embarrassment, anguish, anxiety, and nervousness." It charged PG&E with responsibility for the destruction of Paradise and the deaths of at least forty-two people. The lawsuit laid out wildfires and gas accidents caused by PG&E over decades—two deaths in a 1992 explosion in Santa Rosa, another two in the 2015 Butte Fire, for example—and said the utility demonstrated a "pattern of placing its own profits before the safety of the California residents it serves." It called PG&E's safety record an "abomination."

Defenders of the utility were few. Instead of "Are you suing?" people on the Ridge started to ask one another, "Who's your lawyer?"

For Skye Sedwick, it was a matter of principle. Money from a settlement would help, but a lawsuit was to ensure PG&E answered for what it had done. The utility sealed her resolve when it announced in January filings that it wanted to award $130 million in bonuses to thousands of employees for their performance in 2018. It would later reverse the decision, but Skye had already made up her mind. "That was such a slap in our face," she said. "My dad always paid his PG&E bill on time. And he wasn't even afforded a glimmer of respect."

Even for those with insurance—the majority of homeowners have it but are underinsured, while renters tend not to have it— suing was the self-evident course of action. Town councilman

Mike Zuccolillo had coverage that would help pay for the cost of commercial real estate he owned. He was also underinsured, however, and, as a result of the fire, there were no tenants to occupy those buildings. He was annoyed with PG&E, and saw the condition of its grid as a symptom of a broader problem. "We as a nation don't value maintenance, we don't value infrastructure, we don't value getting ahead of it," he said. "We don't really do anything until there's a serious problem."

The councilman encouraged his constituents to sue as well, telling them they had everything to gain and nothing to lose.

"Part of me thinks PG&E needs to be taught a lesson," he said. "This disaster was totally preventable."

At the time of the Camp Fire, PG&E was already facing intense legal scrutiny. Since 2017, the utility had been under federal supervision for criminal violations of federal pipeline safety regulations in the San Bruno gas pipeline explosion. Because corporations can't be sent to prison, the company was put on probation for five years: it had a court-imposed monitor to oversee its safety performance and had to spend up to $3 million to let the public know of its criminal behavior via television commercials and print ads. It was also required not to violate any local, state, or federal laws.

PG&E's probation was overseen by William Haskell Alsup, a seventy-three-year-old federal judge who was appointed to the court by President Bill Clinton. A gray-haired Harvard graduate originally from Mississippi, he was enamored of California and spent time exploring the Sierra Nevada. A profile by the Federal Bar Association described him as a no-nonsense judge who started his workday at 5:30 a.m. and believed that "justice delayed is justice denied." He was known for his dedication to understanding complex issues. In the landmark *Oracle v. Google* case, in which Oracle accused Google of infringing on its copyright by using thousands of lines of Java code in the Android operating system, Alsup, who had casually studied coding for decades, learned some Java.

Alsup said his role in overseeing PG&E's probation was to "protect the public from further wrongs" and promote the utility's "rehabilitation."

On November 27, two days after authorities announced that the Camp Fire was fully contained, Alsup, in a court filing, directed the utility to explain any role it might have had in that and other recent blazes.

The utility was summoned before Alsup on January 30 for a matter not directly related to the Camp Fire—it was charged with violating its probation by failing to notify officials that it had struck a deal with Butte County for its role in the 2017 Honey Fire, which burned 150 acres. But that court date became a public reckoning over PG&E's culpability in devastating blazes like the one that destroyed Paradise.

Dozens of people, including regulators, fire attorneys, and Cal Fire officials, streamed into a San Francisco courtroom and an overflow space in the largest federal building west of the Mississippi. In attendance, too, was PG&E's interim CEO, John Simon, an attorney who had been with the company in various senior positions since 2007 and sat in the front row. Simon had recently been appointed to replace Geisha Williams, who became the first Latina chief executive of a Fortune 500 company when she took the job in 2017. But she left just ahead of the bankruptcy filing, and was now facing criticism for the $2.5 million severance package she received.

Alsup couldn't put anyone from PG&E in jail, but he could use the bench as a bully pulpit, taking the utility to task and demanding answers. He began by condemning the utility for not taking responsibility for what it had done.

"Every time I ask you to admit something, you say you're still under investigation," Alsup said to a panel of PG&E attorneys and representatives. "Usually a criminal on probation is forthcoming and admits what they need to admit. You haven't admitted much. In fact, very little. I'm going to give you some opportunities, but to my mind there is one very clear-cut pattern here: That PG&E is starting these fires."

The judge was outraged by a recent report from Cal Fire that found PG&E had sparked seventeen major wildfires in 2017. He acknowledged that climate change played a role, as did drought, but said: "The drought did not start the fire. PG&E, according to Cal Fire, started the fire. Global warming did not start the fire. According to Cal Fire, PG&E started it, all 17 of them."

This led a frustrated Alsup to ask: "Does the judge just turn a blind eye and say, 'PG&E, continue your business as usual. Kill more people by starting more fires?' I know it's not quite that simple because we've got to have electricity in this state, but can't we have electricity that is delivered safely?"

Though the Camp Fire was brought up by name only a handful of times, what happened in Paradise seemed to hang over the hearing. And Alsup didn't mince words—when he talked about deaths in wildfires he used the phrase "burned alive."

In a matter of minutes, Alsup found PG&E in violation of its probation over the Honey Fire issue. But the judge was not done with the utility. Alsup wanted to know more about why it had rejected his proposal to add additional—and costly—terms to its probation. PG&E had claimed additional tree trimming and inspection requirements, aimed at mitigating wildfires, would necessitate a $75 billion to $150 billion investment and 650,000 full-time employees, and would lead to substantial cost increases for ratepayers. The company said it nevertheless understood the judge's concerns and was dedicated to reducing risk.

Addressing a PG&E lawyer, Reid Schar, Alsup asked: "Why can't the risk be zero? Why is it that PG&E should be permitted to start a single wildfire?"

Schar replied: "Well, the answer to the first question is bringing the risk to zero is an incredibly complicated series of policy decisions that have to factor in reliability, cost, safety, and there's a tremendous amount of analysis that goes into how best to, for instance, make vegetation management decisions and how aggressive vegetation management should be versus the cost of—"

Alsup interrupted. The firm had paid dividends of $4.5 billion over five years to shareholders, he said.

"Some of that money could have been used to cut trees and trim trees, the very trees that started these 17 fires—or 14 of them I should say. Those trees could have been cut. No. The money went to the shareholders. So why is it PG&E says all the time 'Safety is our number one thing'? I hear it all the time, 'Safety. Safety. Safety,' but it's not really true. Safety is not your number one thing."

The utility insisted that it could not completely eliminate the risks. "We have an inherently dangerous product is the fact," said PG&E attorney Kevin Orsini. "We have electric [sic] running through high-power lines in areas that are incredibly susceptible to wildfire conditions."

Before going into recess, Alsup left PG&E with pointed words. "The last two years of wildfires have been the single-worst catastrophic events that I've seen," he said. The people of the Paradise Ridge and others had died "under horrible conditions.

"In two years, three percent of California burned up. Think about that. Three percent of the whole state burned up. We cannot continue to sustain these kinds of catastrophic injuries to the state, death and destruction. PG&E is not the only source of these fires, but it is a source, and it's really to most of us unthinkable that a public utility would be out there causing that kind of damage."

Alsup turned his attention to Simon, the interim CEO. "I know it's not easy sitting there and listening about all this about your company. I know it's hard. So thank you for being a good citizen and being here to hear this out. I appreciate your coming today."

"Thank you, your Honor," Simon said. "I've listened carefully and I'll take your points back."

In April, droves of people descended a dusty dirt track that led from Magalia's renowned nineteenth-century church to a fenced, earthen ballfield. A few wore Wild West period dress: hoop skirts, beribboned hats, waistcoats, and neckties. The sun was so hot, the sky such a hard and unrelenting blue, that the downpours of previous months seemed an impossible memory.

Sweaty, smiling men followed by piebald and brown donkeys entered the ring, the last stage of a sixty-year-old race called the Donkey Derby. That day they had started down in the valley and proceeded up past waypoints with names like Gravel Gerti's Corner, Dead Man's Curve, and Point of No Return, and in front of crowds at the ballfield they had to complete an obstacle course. The winner was a brown-and-white striver called Poppy and her handler, Henry Schleiger.

This was the first big event of Gold Nugget Days, an annual Paradise Ridge celebration that began in 1959, the hundred-year anniversary of the discovery of the gigantic Dogtown Nugget. Ordinarily the next event would be a parade through downtown Paradise, but this year it was abbreviated, only covering a distance of about 750 feet. Before sparse crowds, two women on horseback carried the US and California flags, followed by a jaunty brown horse with a swishing tail whose rider made it prance down the roadway. A truck pulled a wooden model of the old Magalia church. As many spectators put their hands over their ears, the Devil Mountain Brigade drill team fired shotguns into the air.

Afterward people drove to the Paradise recreation center, magically untouched, where dogwoods flowered in a sea of technicolor grass. Locals set up stalls selling clothes, plants, and honey. People kept sneezing because there was so much pollen in the air.

Town councilman Mike Zuccolillo thought it almost looked like the fire had never happened. Of course, the homes across the street were destroyed, and behind the fair, a church's parish hall stood in ruin. But you could try not to notice them. "If you live in San Francisco you have earthquakes and no one tells you can't build there," he said. "We have to rebuild." Zuccolillo himself was planning to.

Events like Gold Nugget Days were as important for getting Paradise going again as opening its schools, said Rick Silva, editor of the *Paradise Post*, which had published without pause since the disaster. "If you don't have it then you lose the momentum for next year," he said as he surveyed the scene in the park. He estimated there were only one-third as many attendees as usual. "You gotta start it, gotta keep it going."

"I wish I would've grabbed my home videos, my jewelry," a woman told friends she'd bumped into. She was wearing a pin with the words "Moving Forward."

"If I'd just grabbed those I could have dealt with everything else," she said. "All your history. All your life. Everybody's wedding. My kids when they were young." She added: "One thing about not having a lot of furniture, you don't have to dust very much."

By 3:00 p.m. the fair in the park was becoming deserted, despite the fact that it wasn't scheduled to end for another hour.

The idea of rebuilding was baked in from the very beginning. Five days after the fire, city manager Lauren Gill had suggested at a council meeting that Paradise would be "a great place to live again."

Jason and Meagann Buzzard went to Paradise Town Hall in March to receive their rebuild permit, the first issued by the town since the fire.

Living in a trailer in Chico with their eleven-year-old daughter, Grace, rebuilding had never been in question for them. Three days after the fire, Meagann began calling contractors. "The time for mourning the loss of our house and all our stuff is over and it's time to move forward," Jason said. Both he and his wife were born and raised in Paradise—he was a realtor, Meagann was in marketing—and he wanted to be part of the new, younger vanguard that he thought would re-create the town. He wasn't afraid of another fire—they were just a fact of life. "Every year we have this fire that starts in the Concow area and every year we see the smoke," he said. In his view, the destruction of Paradise was a freak occurrence.

By the fall of 2019, the town had issued about 250 rebuild permits. To spur reconstruction efforts, a Pittsburgh-based firm called Urban Design Associates facilitated seven community meetings and "listening sessions" to produce a town recovery plan. In the architectural renderings of a picture-perfect small community, the Paradise of the future would have a walkable downtown and even a live-work space for entrepreneurs. The blueprint comprised almost forty discrete projects, the highest priority of which included several related to fire safety: a better alert system, better evacuation routes, more brush trimming. They were short on specifics and more a statement of purpose.

This was going to be a different sort of Paradise, of necessity more rule-bound, as well as less accessible to those with emptier pockets. "Much of the housing in Paradise prior to the fire was naturally occurring affordable housing," the plan admitted.

"Given the cost of construction, it will be hard to replicate the levels of housing affordability."

Residents wondered if Paradise, with its upgraded homes and utilities, would become a wealthier community populated only by those with the resources to rebuild and pay for the growing cost of homeowners' insurance, a requirement for anyone with a mortgage. Others were dropped by insurers who deemed the Ridge too risky for coverage.

When people are displaced due to natural disaster, economic and cultural shifts may hinder their efforts to return. After Hurricane Katrina in 2005, a study found that areas of New Orleans with worse storm damage were more likely to have gentrified a decade later, with significant increase in housing prices and more college-educated residents. There were no clear answers as to why, but the authors speculated that the most impacted neighborhoods may have required extensive infrastructure redevelopment, paving the way for private investors to reap the benefits.

A forecast by the county found that Paradise might reach its former population by 2040, but only in the most optimistic of scenarios. Yet Paradise's elders and observers wondered whether this was even something it was wise to aim for. "Should Paradise rebuild?" said Ken Pimlott, the former head of Cal Fire, several months after the blaze. "That's the $64 million question."

Pimlott had made waves immediately following the fire when he seemed to suggest in an interview that it was simply too dangerous to allow new construction in California's wildlands. By contrast, California governor Gavin Newsom would tell an interviewer that "there's something that is truly Californian about the wilderness and the wild and pioneering spirit." Frequently towns like Paradise were the only places in California still affordable to a person on disability or the minimum wage. Later Pimlott clarified that he had been misinterpreted—he meant building should only be permitted if the wildfire risks could be mitigated.

The urge to go back made sense to him. "I get it," he said. He was putting up a home of his own in rural El Dorado County, in the Sierra foothills. "I want my piece of California that I've invested in for a long time. But also I understand the risk."

The practical steps Pimlott was taking were the kinds of things that the future architects of Paradise would have to incorporate, and then some. His property, 71 acres of oak and mixed conifer woodland, had actually burned in a wildfire a few months before he bought it. "I guarantee there will be another fire out there," he said. "We watch it every day."

His home would be built to the baseline code requirements of any new home in California's at-risk areas—ventilation openings would be covered to ensure embers could not enter and roof and wall materials would be nonflammable. He would cluster structures together to make them easier to protect, and he would keep vegetation down to ensure 100 feet or more of "defensible space" around them. Seeing as there were no hydrants, he was creating a water reservoir. There would be two ways in and two ways out to provide safe and easy access to firefighters. Pimlott's approach to living in the wilderness was anything but casual. "I don't think it's an undertaking for the faint of heart," he said.

It went without saying that the new Paradise would need a better way of alerting people to disaster than its opt-in telephone alert system, which many had not signed up for and which relied on fragile networks of cell towers. Paradise chose this approach, said town emergency planner Jim Broshears, because the other option was a technology that could ping every phone in a given area, but was not capable of much geographic precision, and could not target residents in specific evacuation zones. Broshears and others feared that undiscriminating warnings would cause mass panic and gridlock—though that was what happened in the Camp Fire anyway. No future warning system should rely on voluntary signup, he said. But figuring out a better option was the responsibility of the state of California, he added, not a small town with limited resources.

Then there was the matter of an evacuation strategy. The Camp Fire showed that planners did not think apocalyptically enough, positing only the evacuation of people in increments, based on the zones in which they lived, as opposed to the departure of every resident at once. "Who would have dreamed that you would have a fire that cut off Skyway to north and Skyway to south in the same fire?" said Broshears. "We didn't, to our detriment."

In fact, no town or city in the United States could immediately evacuate every single resident at once, said Mike Robinson, who heads a team that simulates and studies disaster evacuations at Old Dominion University in Norfolk, Virginia. First, every person must be ready to act on an evacuation order as soon as it is received—an unrealistic expectation. Then "it's like leaving a major sporting event. There's going to be massive congestion for a while," he said. "And now you add smoke, fires and cars burning and people frightened for their lives, and it's not going to work."

Paradise was an order of devastation greater than anything Robinson had ever seen, and he was not inclined to finger emergency planners as the main culprits. "It's not fair to blame them for everybody not getting out when this is the new world. Nobody has ever seen fires like we've had in the past two years. When you're the first one it happens to, it goes poorly." Instead of mass evacuations, he proposed, communities should create refuge areas where people could ride a fire out. There were a few in Paradise, but they were inaccessible to many evacuees.

Considering Paradise's precarious location, Broshears said, perhaps its size should be limited. "I don't believe that 27,000 people should have been living in the town of Paradise," he said. "Now that we have a chance to do the planning right, I don't think we should ever again have that many people." The head of a local fire safety council, Phil John, ventured an even more searching question. "Part of me wonders, too, should the town of Paradise even exist anymore?" Its population had shrunk so much, perhaps it should unincorporate, jettisoning its mayor and town council and police force, and go back to being under the jurisdiction of the county, as it had been before 1979.

Thousands of Camp Fire victims ended up suing PG&E. As a result of these and other claims, PG&E filed Chapter 11 bankruptcy two and a half months after the fire, its second time in eighteen years. The company listed about $71 billion in assets and nearly $52 billion in debts, and cited $30 billion or more in potential liabilities and lawsuits from victims of the 2017 and 2018 wildfires. Bankruptcy would allow the company to continue operations as it sought to resolve its financial issues, and would not relieve it of responsibility to victims, though any payouts might be smaller than they would have been otherwise.

In public, PG&E defended itself by arguing that though it was responsible for destructive fires, there was only so much it could do as a result of the changing climate and development in more remote areas. It claimed there weren't enough qualified tree trimmers in the entire United States to do the souped-up maintenance work being asked of it.

"California has been in some of the worst drought history it's ever faced," one PG&E lineman said. Referring to the Camp Fire, he added: "It didn't matter what happened that day with the winds as bad as they were. It could have been anything. I have never seen them dodge trying to do maintenance.

"If there's 70 mile an hour winds that's an act of God. You couldn't build anything to withstand that."

In February, one month after declaring bankruptcy, PG&E acknowledged in an earnings report that it was "probable" that its equipment would be determined to be "an ignition point" of California's deadliest fire.

Cal Fire investigators concurred. "After a very meticulous and thorough investigation, CAL FIRE has determined that the Camp Fire was caused by electrical transmission lines owned and operated by Pacific Gas and Electricity," it announced in a statement in May.

"The fire started in the early morning hours near the community of Pulga in Butte County. The tinder dry vegetation and Red Flag conditions consisting of strong winds, low humidity and warm temperatures promoted this fire and caused extreme rates of spread." It added that there was a second ignition point about four miles from the first. This blaze was caused by vegetation coming into contact with PG&E power lines, but it was consumed by the first fire that had broken out near Camp Creek Road.

That same month, Judge Alsup ordered company brass to tour Paradise and the San Bruno neighborhood affected by the 2010 pipeline explosion "to gain a firsthand understanding of the harm inflicted on those communities," as part of the sentence for its probation violation. In early June, Alsup joined the executives, including the new CEO Bill Johnson, who had already visited the Paradise area in May and met with PG&E workers who were helping rebuild gas and electric infrastructure, and some of whom had lost their own homes. Alsup's group saw the town performing arts center, the closed Feather River hospital, and the remnants of Paradise Elementary School. The visit was kept quiet. Officials didn't want to create a spectacle, and few in the town even knew it was taking place, including councilman Zuccolillo. He hoped their visit was difficult.

"I want this tour to be miserable," Zuccolillo told media. "This is not going to the zoo to pet a giraffe. I want people screaming and yelling. I want them to feel that they destroyed our town."

Behind the scenes, Paradise, Butte County, and other towns and areas affected by PG&E-sparked fires had been in negotiation with the utility for months.

Eleven days after the visit, PG&E settled with several cities and counties over the 2015, 2017, and 2018 wildfires. A sum of $415 million would be split between nine Wine Country cities and counties impacted by the 2017 fires. Paradise received $270 million, its parks district received $47.5 million, and Butte County was awarded $252 million. The settlements came more quickly than anyone expected, and individuals later accepted a payout of $13.5 billion, which was intended to cover the Camp Fire and various other conflagrations.

The payout to Paradise guaranteed it a future. Its tax base had shrunk by 90 percent and sales-tax revenue dramatically declined. Now it would be able to complete costly projects, such as establishing a sewer system at a possible cost of tens of millions of dollars.

Despite settling its financial responsibilities, the utility would still have to explain how its actions, or inaction, contributed to the deaths of eighty-five people in the Camp Fire. Judge Alsup had read a report in the *Wall Street Journal* in July that described how the company had been aware that high-voltage power lines in its 18,500-mile transmission system were "dangerously outdated," yet failed to upgrade them.

Approximately fifty of the steel towers on the Caribou-Palermo line were in need of replacement, the newspaper reported. The mean life expectancy of transmission towers was sixty-five years; PG&E estimated that the average age of its towers was sixty-eight, and the oldest were a hundred and eight. The Caribou-Palermo line is part of a section of the grid so old that portions were once candidates for inclusion on the National Register of Historic Places. It had delayed safety work on the line for more than five years.

As if addressing a recalcitrant schoolchild, Judge Alsup ordered PG&E to respond to the article on a "paragraph-by-paragraph basis" within twenty-one days, in a document no longer than forty double-spaced pages.

In its filing, PG&E disputed that it had neglected maintenance, but ultimately acknowledged the report's central findings that it knew its aging high-voltage transmission lines could fail and cause wildfires. There were actually about sixty towers—more than the *Journal* initially reported—that PG&E had identified for replacement on the Caribou-Palermo transmission line, the utility said. The tower that sparked the fire was not one of them. It had delayed several upgrades to the line, but said that completing such maintenance would not have prevented the Camp Fire. The purpose of the planned upgrades "was not to identify and fix worn or broken parts, such as the hook on the transmission tower that failed and caused the Camp Fire to ignite," but to address design issues: the older towers were unsuitable for extensions to raise their height.

California governor Newsom declared a state of emergency in March in response to the increasing risk of wildfires and in order to expedite dozens of forest management projects, including the creation of fuel breaks near some of the highest risk communities. PG&E undertook inspections of 700,000 distribution poles and 50,000 transmission structures, and it shut down the Caribou-Palermo line. It committed to burying underground all electric distribution power lines in Paradise and some of the surrounding areas at no extra cost, though doing so on a large scale would take decades and tens of billions of dollars, it said. It replaced most members of its board of directors. California lawmakers devised a plan to help pay for the increasing costs of wildfires—and put at least part of the burden on ratepayers. They established a $21 billion fund to help utilities cover damage liabilities, funded by the companies themselves and a monthly fee charged to customers.

Regulators announced bold plans, saying they were looking at whether the utility should be converted into a publicly owned company, divided into regional subsidiaries or split into separate electric and gas companies. San Francisco and other cities expressed interest in taking over portions of the company's network. Steven Weissman, the former judge who helped oversee PG&E, suggested that the problems with the utility were far more profound than most people suspected.

"Unfortunately with the state of their equipment it's hard to imagine they're not going to have more catastrophic fires," he said. "Probably the best thing they can do is cross their fingers and hope they have a lucky decade."

"PG&E as it is now developed over decades—it grew organically. It was, 'Well, people are moving over there, we need to put some lines over there.' These are decisions that are made piecemeal over a century so it's not hard to believe they could lose track of what they've got.

"There is a theory of complexity that some institutions can grow to a level of complexity that no one anywhere understands what's going on. I put an operation the size of PG&E in that category."

In the spring, Paradise bloomed. Daffodils and irises flowered in neat groups in front of bare home foundations. Grasses emerged unchecked and, swaying in the wind, lent the town a prairie feel. The walls of the canyons around Paradise were streaked with wildflowers like the yellow foothill poppy, a fire-follower species that is common in the wake of blazes.

On a glorious afternoon, in a car heading up Skyway to Paradise, Don Hankins, a professor of geography and planning at Chico State, noted all the new plant life that had sprouted since the Camp Fire. There were invasive species, like the sunshiny St. John's wort and nodding wild oat, that tended to be quicker to reestablish themselves than natives. In the valleys Hankins had made peculiar finds of plants that had presumably washed down as seeds from burnt homes, including corn and melon. But he also came across intriguing natives, such as some varieties of the nightshade family. Everywhere there were black oaks whose bark was burned smooth but showed signs of regenerating, putting out new stems from the base or from buds high in the sky that had previously been dormant. Frass on the bark showed that insects were busy burrowing away inside.

Hankins's grandfathers were both Native American, and he had devoted himself to studying the effects of fire on a landscape that native peoples once let burn freely. He was the only remaining speaker of the Plains Miwok language, which he learned from its last native speaker before she passed away in the 1990s. He led the way to where Skyway traversed the high isthmus of rock linking Paradise and Magalia. There, on a spectacular overlook where the land began its precipitous eastward drop to the canyon floor, the fire had done its work and had left the gnarled trunks of a few dozen blackened, apparently completely dead trees.

But something was not right about the rocky ground beneath them. The exposed stones seemed to be faintly glowing.

This was a band of serpentine rock jutting from the earth, its color a characteristic, alluring pale green. Serpentine is associated with gold, and old mine shafts could be found all over the Ridge alongside spurs of this mineral. The soils produced atop serpentine are inhospitable to many plants because they are low in nutrients and high in heavy metals, yet the few that could cope with it were scattered about with blossoms that seemed particularly jewel-like on their jade bed, such as California brodiaeas, woolly sunflowers, and the toyon shrub. Most remarkable of all were the MacNab cypresses, a conifer endemic to California, with reddish branches and sprays of feathery leaves.

On this particular spur of land, seemingly all the trees had been killed by the fire. But Hankins, taking care not to trample on exposed soil and skipping from boulder to boulder, had found something special. The MacNab also just happens to be propagated by fire. Its seed-bearing cones can only be cracked open by extreme heat. And across this serpentine plain there were innumerable tufted seedlings poking up from the ground.

"At first I thought the timing of this fire was bad and it would lead to nothing," said Hankins. He was referring to the possibility that the fire would favor invasives over local species. "This is clearly not nothing."

For months after the fire, Iris Natividad felt a pull to Paradise. "I would go to the property and to me it would feel like I needed to be there, go there, be close to Andrew because that's where he passed away." But then, one day in June, she went to the lot and found that it had been cleared. Instead of the molten mass of her and Downer's life, there was merely a blank expanse of packed earth scored lightly with machinery tire marks. A few toothpick pines had been left, as had a dead-looking oak that stood outside their front door. On one of the trees she found that someone had inscribed a large heart into the charcoal bark.

"It doesn't even look like I live here," she said. "It's like everything has been erased—like all the memories are gone, everything gone. Even in the devastation, at least you saw parts of your home." In a text message that accompanied a photo of the lot, she wrote: "No reason to go back."

And yet, a week or two later, she sent another photograph: of the roadside billboard she had helped Treasures from Paradise procure on Skyway. "I feel like I want to help my people in Paradise," she said. "But then there's a tug that feels like you need to move on—you need to stop thinking about Paradise and you need to move on."

It did not help that she continued to live in a high-risk fire zone, in the community of Bangor. That summer and fall, PG&E dramatically expanded its preventive power shutoffs.

"Our lives now consist of the new normals of evacuations, of power being turned off by PG&E. I haven't had power in Bangor for four days—that's our new normal," she said. "It's been a constant trauma for a lot of people because it's a constant reminder of what happened to us."

Skye Sedwick felt drawn to the Ridge, too. Under a warm June sun, she sifted through the ashes of her old life, carefully navigating over the rusted nails, broken glass, and unexpected craters in search of copper electrical wiring. Her light jeans grayed and her fingernails darkened as she sorted through the ruins. Jude trailed behind her, getting tangled under fallen power lines, before he thought better of it and sat on grass freshly sprouted from the hillside.

There was little left to salvage from the cabin Skye's family had lived in for the better part of a century. But she recycled some of the scrap for income, and she could sell any of the wiring she found. It would give her a bit of cash to keep gas in her car, or afford a lunch out every now and then. Besides, if she didn't take it, a looter would, she thought.

Skye hadn't yet spent much time at the property. It was hard to get to, and hard to stay there. Even when she did manage it, she worried about the toxins she was inhaling, and it was difficult for her to do much actual cleanup with her bad back. But on this day, she was determined to work, pain from her degenerative disc disease be damned. Every now and then she came across things that made her smile, like the shard of a mug that had once belonged to her father's third wife, the love of his life.

Slowly Skye had begun to put the pieces of her father's last day together. Parts of the narrative were provided by Bubba Shipman, the man who thought Sedwick might be an angel. He'd only realized what had happened to Sedwick when he heard his name on the radio, and after bumping into Skye at a fire survivor resource center he told her about the time he spent with her father. The experience bonded them, and Shipman tried to help Skye where he could, offering her a cleaning job when she needed it. He wanted to start up another volunteer fire station, and he planned to repair the headstone that his cedar had cracked in the cemetery.

On that fine June day, another key player in Sedwick's story stood one hundred yards from the site of Sedwick's house: firefighter Ray Johnson. He had talked to Skye on the phone once in the spring after her lawyer connected them, and it brought them both some closure. Yet they had never met in person, and neither knew that the other was paying a visit to the Ridge. Johnson was in the parking lot of what had once been the Depot, where it was hard to tell the restaurant had ever existed. For months, the foundation had been covered with warped metal and piles of dust that turned mushier with every rainstorm. Even the partially melted hoses the firefighters had cut, or that had torn from the trucks as they fled, remained, a memorial to all that occurred. But by mid-June the county cleared it and there was only a neat, level pile of bronze-colored dirt where the Depot had stood.

The wind was fierce, rocking the trees across the street in the same way it had the night of the fire. Two were so badly burned they had been reduced to withered trunks. They were dead long before the fire, and now they were a hazard. *They're getting ready to fall over,* Johnson said. *They could fall today.*

There was a flash of auburn hair at the top of the hillside—Sedwick's property. Johnson hurried up the driveway.

Skye cried when she realized who it was, bringing her hands to her cheeks. Stepping over a pile of rubble and wiring, they hugged, Skye and the man who last saw her father alive.

A small smile forming on her face, Skye told Johnson: "I'm a mess."

"It's fine," he said gently. "You can be a mess. Your dad passed away."

For a few minutes, they chatted about Magalia, and an attic fire Johnson had once fought next door, until Skye asked him how it had happened, and where her father had passed.

"John was just walking up the driveway," Johnson said. "It was just full fire all around."

Of everything that was lost, the one thing Skye wanted to find was her father's wedding band. Sedwick was never someone who cared much about material things. But it had been important to him, even more so after his wife died. Johnson told her that if the ring was with the body, the authorities would have collected it. Contacting the coroner would be another task for Skye to add to her to-do list.

They discussed rebuilding. Johnson had lived on the Ridge his whole life and had nowhere else to go. "We can't afford anything," he said. "I can build a nicer house here than I can buy used in Chico."

Skye, too, was on the verge of deciding to come back. She was never going to sell the land that she and her dad had loved so much. The floorplan would be as it was before, right down to the bathtub on the porch. But there was one nonnegotiable update. It would have to be fireproof.

For those who had devoted their careers to protecting the communities of the Paradise Ridge, the question of rebuilding was tinged with bitterness. Chris Haile, the former Cal Fire division chief, had observed the fire developing on November 8 and sent Facebook messages to residents, and he seemed as if he couldn't quite believe all the pieces of the awful puzzle that had slotted so neatly together. First there was the wildland town on its overlook—an unlucky spot because heat rises, and fire wants to go up. Then there was the raging north wind. Not least there was the fire season that stretched on well past when it used to.

"What happened in the Camp Fire is the fire that professionals knew was going to happen, but I don't think anyone foresaw a whole town wiped off the map," he said. "This was the perfect fire."

It wasn't his hometown, but the loss was personal. "I spent six years up there protecting that place and I went up there and it was obliterated."

His eyes reddened and his voice quavered.

"It brings me to tears. We spent all that time up there, all that planning, especially with Magalia. All those drills. Mother Nature came in and showed us how big her balls were. Sometimes you get the bear and sometimes the bear gets you."

The risk was so obvious, he said. So many blazes had burned in the Camp Fire's footprint in the preceding decades.

"If I'm thinking as a firefighter, I'm thinking I sure as hell wouldn't want to live there," he said. "It happened. It finally happened.

"You bet it can happen again. Lightning can strike twice."

Epilogue

In the wake of the Camp Fire, a century of certainties about the ability of humans to dominate fire were in question.

To Lincoln Bramwell, the chief historian at the US Forest Service, the story of Paradise "reads like these accounts from the late 19th century," back before wildfire had been brought under paramilitary command. In October 1871, for instance, after railway workers sparked a brush fire, the city of Peshtigo, in Wisconsin, was consumed in the span of an hour, and 1,500 people died there and elsewhere across the 1.2 million acres that were scorched. "I see us going back to the future," Bramwell said. "Going back to a time when fire was not in our control."

For some researchers, it no longer seemed rash to think that a wildfire could, under the right conditions, maraud into the very core of major metropolitan areas, such as Los Angeles, San Diego, and the cities of the San Francisco Bay Area. In 1991, dozens had been killed on the urban fringe in Oakland. "Where I live in the flatlands of Berkeley," a comprehensively built-up and urbane place, "as the crow flies it's less than a mile from where the Oakland Hills Fire started," said Faith Kearns, a University of California scientist with a specialism in wildfire. It wasn't completely

farfetched to her that a blaze could reach her home. "My neighborhood is full of Victorians. My neighbor's window is about six
feet away from my own—" she said, pausing in thought. "I think
we'll see it. I think we'll see it."

From his ranch in the Central Valley town of Williams, former
California governor Jerry Brown had seen many of California's
fires in the guise of smoke that hung overhead. Brown served two
terms as California governor starting in 1975, then was reelected
for another two terms beginning in 2011. By the end of a political career that spanned the state's worst drought in 1,200 years
and such a parade of fires that they blurred into one another, he
was clear-eyed about where America was headed. "Paradise was a
horror," he said. "With climate change we can expect a lot more."

In fact, the town quickly became a poster child for the climate
crisis. Bernie Sanders unveiled a $16 trillion climate plan for his
2020 presidential run on a visit to Paradise. The environmental
activist Greta Thunberg toured it in October 2019.

That fall, it felt like déjà vu. Millions of people across the state
were forced to go without power for days as PG&E and other
utilities de-energized electric distribution lines during dry and
windy conditions. It seemed unfathomable that in twenty-first-
century America, so many would have to live in the dark for as
long as a week, with refrigerated food turning rotten and dead
medical devices. But, of course, the loss of an entire town in a
wildfire had also been unfathomable.

Around the same time, PG&E appeared to have started yet
another massive blaze, this time in Sonoma County, in an area
in which it had said it was shutting off the power. The utility
reported to regulators that it had an equipment malfunction on a
transmission line that remained in use. No one died in the Kincade Fire, in part because authorities ordered an evacuation of up
to 200,000 people, some of whom had just moved into homes
rebuilt after fires that occurred in 2017.

At a press conference, California governor Gavin Newsom addressed the chief suspect. "It took us decades to get here, but we will get out of this mess. We will hold them to an account that they have never been held in the past. We will do everything in our power to restructure PG&E so it is a completely different entity."

His remarks were full of resolve. They were also familiar. Californians had heard these sentiments before, to no avail. And on and around the ruined Paradise Ridge, people watched the fires with a familiar sense of dread.

ACKNOWLEDGMENTS

Our deepest thanks go to the residents of the Paradise Ridge and surrounding towns. You told us your stories and welcomed us into your temporary homes and FEMA trailers and onto your burned-out lots despite the grief and tragedy that had engulfed you. We are honored that you trusted us with your experiences and your memories.

When the fire broke out in Paradise on a Thursday morning, we both watched it unfold on social media. Within hours, we could smell the smoke from the *Guardian*'s office in Oakland. For one of us, the story was more personal than most. Dani lived just 20 minutes down the road from Paradise for a decade, until coming to work for the *Guardian* in summer 2018. She fielded worried texts from family who didn't know the fate of their town or home.

Within hours, Dani was dispatched to cover the fire. Having overseen coverage of previous California blazes, Alastair edited some of Dani's first stories. We began collaborating as co-authors a few weeks later, and are indebted to Malik Meer, Eline Gordts, Mark Oliver, and Charlotte Simmonds for wisely shepherding our dispatches. (Some of the reporting in this book first appeared in the *Guardian*, and was funded in part by a grant from the Society of Environmental Journalists.) This project would not have been possible without the support of our beloved West Coast bureau, which punches far above its weight. We are also immensely grateful for the backing of the broader *Guardian* team, including John Mulholland, Jane Spencer, and Rachel White, and are proud to be part of it.

Our editor at Norton, Matt Weiland, recognized the potential of this project, and guided us through the process with the kind of unfailing encouragement and deft editing of which writers dream. We are lucky to have him and the rest of Norton on our side. Zoë Pagnamenta, our agent, has been an invaluable guide to the unfamiliar world of book publishing. Our researcher, Margaret Katcher, helped us understand the Camp Fire's intricate backstory, and our fact-checker, Sameen Gauhar, saved our bacon.

From Alastair: I was brought up in Manchester, England, by parents who believed I was capable of whatever I set my mind to, and my first thanks are to them, Marsha and Colin Gee, as well as to my siblings, grandparents, and extended family. After college, I did not know what to do with myself, except that I wanted to write and to live abroad. I went to Russia for a one-month language course at Moscow State University, and while I was there I was offered a copy-editing job at *The Moscow Times* by the felicitously named copy chief, Emily Gee.

Like many others, I learned how to be a journalist at the paper, a surprisingly hard-hitting, American-style daily with an international readership. I became a reporter and an editor, and owed a great deal to colleagues such as Peter Savodnik, who blew into the newsroom one day from Washington, D.C., and whose career advice I still hold to. We all need mentors such as these, whose vision extends further than our own and who help us to see what is possible. After quitting to write for US and UK publications, I depended on the fellowship of other freelance journalists such as Marina Kamenev and Joyce Man, through the stresses of being self-employed, the idiosyncrasies of post-Soviet life, and even an arrest by Russian police. One month became four years.

I am fortunate that I landed next in San Francisco, because it is home to the Writers Grotto, a collective that welcomed me to its downtown workspace and enfolded me in its community. I wrote about the environment, science, and culture for outlets such as the *Economist*, the *New Yorker* online, and the *New York Times*. These stories took me deep into the Mojave Desert and high into the Sierra Nevada, kindled my love of the California wilderness and educated me about the challenges it faces.

Eventually I was hired by two superb *Guardian* editors, Paul Lewis and Merope Mills, to be America's only homelessness editor and report on the West Coast housing crisis. I branched out, writing and commissioning stories about US public lands and helming some of our wildfire coverage. The *Guardian* allowed me to follow my passions (I now edit a number of our special projects in the US). It also introduced me to Dani, whose grit, enthusiasm, and professionalism are a thing to behold. I feel as though I won the co-author lottery.

I share my adoration of California, its plants and landscapes and out-of-the-way towns, with my husband, Aish Shukla. His joy and creativity light my days. There is no more I could have asked for from him on this journey, no better partner. We got married in a redwood grove two months before the Camp Fire.

From Dani: Paradise was a community I knew and loved. Butte County, and small towns like Paradise, are where I feel most at home in the world.

I covered the community during the three years I spent as a

reporter for the *Chico Enterprise-Record*, writing about the details of foothill life: the annual opening of the ice skating rink, a biker toy drive, and the passing of a beloved high school teacher. Friends and family worked and lived there. When Paradise burned, my cousin and her father lost their home. They were unharmed, but three people in their neighborhood died, including the man who lived just next door to them. I felt a deep responsibility to tell the town's story.

The time I spent at California State University, Chico, and with its student newspaper, the *Orion*, established the foundation of my career, and I built on it with a graduate degree from the Medill journalism school at Northwestern University.

The *Enterprise-Record* hired me after I finished school, and gave me the freedom to pursue whatever stories I wanted, as well as, most importantly, the guidance and skills to do them well. The team I worked with for three years at the *Enterprise-Record*, along with the *Oroville Mercury-Register* and *Paradise Post*, shaped me into the journalist I am today. My editors, David Little, Steve Schoonover, and Tang Lor, and fellow reporters—particularly my deskmates, Risa Johnson, Ashiah Scharaga, and Andre Byik— set an example of how to do impactful, vital reporting, even with diminishing resources, a circumstance all too common in local news.

After leaving the *Enterprise-Record*, I joined the *Guardian*'s vibrant West Coast office, filled with a small group of mighty, unflappable editors, copy editors, and reporters. The *Guardian* offered me the opportunity to bring the story of the fire, and its impact on the communities I know so well, to the world. My colleagues there have been steadfast champions of my reporting, inspiring me each day, and I am grateful to work among them. It was because of the *Guardian* I got the opportunity to work with Alastair, who has been the best writing partner I could have asked for, as well as a thoughtful mentor and friend. His dedication to this story and tireless nature have been invaluable to me.

I owe profound thanks to my parents, David and Angie Anguiano, for their love and unwavering belief in me—they have always been my biggest cheerleaders—as well as to my second parents, Nena Anguiano and Joaquin Moreno, who taught me to fearlessly follow my dreams, and generously hosted me and offered many a hot meal during my coverage of the fire and reporting for the book. My partner, Shane Price, was endlessly supportive and patient as I spent every spare minute in Paradise, canceling countless movie dates, and was an excellent solo parent to our cat, Moby.

My tías and tíos, cousins, siblings, and friends urged me forward throughout this journey, providing food and wine and always an ear to listen. My cousin Isabella Santiago guided me through Paradise and the ruins of her house, offering invaluable insight on the impact of losing one's community, home, and almost every belonging, and helped me in more ways than I can count.

I am eternally grateful to my maternal grandmother, Arlene Key, who encouraged my love of reading and writing from a young age, and my paternal grandparents, Elena and Jose Anguiano, who bravely left Mexico to come to the US more than sixty years ago. Their love, and long days of toiling as migrant farm laborers, helped to build lives of unimaginable opportunity for their children and grandchildren.

NOTES ON SOURCES

Beginning immediately after the fire as reporters for the *Guardian*, and over the course of a year as authors of this book, we interviewed hundreds of residents of the Paradise Ridge and surrounding areas, and friends and relatives of those who lived there, as well as first responders. The book is grounded in the stories they graciously told us despite the enormous hardships many were facing. We relied in particular on our interviews with Skye Sedwick; Simona MacAngus; Iris Natividad; Angela Loo; James Burt; Red Remley; Lonnie Walker; Pat and Tom LeBlanc; Anna Dise; Christina Taft; Chellie Saleen; Rachelle Sanders; Bill Goggia; Arthur Villa; Sadia Quint; Joe Kennedy; Elliot Hopkins; Jeff Edson; Mike Nichols; Chris Haile; Matt McKenzie; Jeff Evans; Allyn Pierce; Chelsea West; Jessee Ernest; Matt Strausbaugh; Rob Nichols; Waylon Shipman; Tammie Konicki; Pete and Peggy Moak; Renee and Scott Carlin; Lita Siebenthal; Jim Broshears; Lauren Gill; Jody Jones; Mike Zuccolillo; Stephen Murray; Eric Reinbold; David Akin; Anthony Borgman; Kassidy Honea; Steve Bertagna; Kyle Schukei; Ray Johnson; Lloyd Romine; Andrew Duran; David Micalizio; Loren Lighthall; Robert Edwards; Marissa Nypl; Sheila Craft; Amber McErquiaga; Liana Paxton; Anastasia Skinner; Aaron Parmley; Megan Brown; Phil John; Calli-Jane DeAnda; Allen Myers; Jaime O'Neill; Matt Chauvin; Patricia Smith; Trisha Wells; French Clements; Shelby Chase; and Luke Hugel.

For matters related to the academic study of wildfire, we interviewed Lincoln Bramwell; Daniel Swain; Greg Asner; Faith Kearns; Craig Clements; Crystal Kolden; Alistair Smith; Scott Stephens; Matt Hurteau; Stephen J. Pyne; Eric Knapp; Chris Dicus; Craig Thomas; Bob Yokelson; Philip Rundel; Jeff Chambers; and Don Hankins. We also spoke with fire incident meteorologists Alex Hoon and Aviva Braun. Regarding PG&E, we are thankful to Steven Weissman and Alexandra von Meier.

Sources that were particularly helpful to individual chapters are noted below.

Prologue

Information on the fire's early speed came from the study *Rapid Remote Sensing Assessment of Landscape-Scale impacts from the California Camp Fire* by Jeffrey Chambers and colleagues.

Chapter 1: A Gold Rush Town

Quoted writings by John Sedwick came from the local-history publication *Tales of the Paradise Ridge*. Details on the history of Paradise came from sources including *Tales, The Golden Ridge, The Golden Ridge 2*, and the *Dogtown Territorial Quarterly*; back issues of the *Paradise Post*, the *Chico Enterprise-Record*, the *San Francisco Examiner,* and *LIFE* magazine; Paradise High School yearbooks; the annual report of the Adventist Health Feather River hospital; *The Flumes and Trails of Paradise: Hiking Through History on the Ridge* by Roger and Helen Ekins; and recollections and writings by Don Criswell, Barbara Hendrickson Ramsay, Linda Lau Anusasananan, Pam Figge, and Dane Cameron. The man who described himself as the Paradise "town drunk" and became the mayor is Steve Culleton. Jaime O'Neill and Matt Chauvin are the neighbors on opposite ends of the political spectrum. Demographic and economic data for Paradise came from the US Census Bureau and Trulia.

For information on California flora we relied heavily on Allan A. Schoenherr's *A Natural History of California,* as well as on Kenneth Starr's *Americans and the California Dream, 1850–1915*; Vinson Brown's *The Sierra Nevadan Wildlife Region: Its Common Wild Animals and Plants*; Verna R. Johnston's *Sierra Nevada: The Naturalist's Companion*; Jared Farmer's *Trees in Paradise: A California History*; and David Carle's *Introduction to Fire in California*; in addition to the National Park Service, the US Forest Service, and the database of ancient trees maintained by Rocky Mountain Tree-Ring Research. Helpful context was provided by John Muir's *My First Summer in the Sierra* and *The Mountains of California*, Clarence King's *Mountaineering in the Sierra Nevada*, and Mark Twain's *Roughing It*.

We gleaned much about Native American history from M. Kat Anderson's

Tending the Wild: Native American Knowledge and the Management of California's Natural Resources; Benjamin Madley's *An American Genocide: The United States and the California Indian Catastrophe, 1846–1873*; the *Handbook of North American Indians*; and Theodora Kroeber's *Ishi in Two Worlds*. California and Gold Rush history came from Starr's *Americans and the California Dream* and *California: A History* and H. W. Brands's *The Age of Gold: The California Gold Rush and the New American Dream*. The history of fire in California is detailed in the paper *Socio-ecological transitions trigger fire regime shifts and modulate fire–climate interactions in the Sierra Nevada, USA, 1600–2015 CE* by Alan H. Taylor, Valerie Trouet, Carl N. Skinner, and Scott Stephens, and *Introduction to Wildland Fire* by Stephen J. Pyne, Patricia L. Andrews, and R. D. Laven, and for climate-change data we used the state publication *Indicators of Climate Change in California*.

The history of America's approach to wildfires came from Norman Maclean's *Young Men and Fire*; Michael Kodas' *Megafire: The Race to Extinguish A Deadly Epidemic of Flame*; Timothy Egan's *The Big Burn: Teddy Roosevelt and the Fire That Saved America*; and three books on the Yarnell Hill Fire (Fernanda Santos's *The Fire Line: The Story of the Granite Mountain Hotshots and One of the Deadliest Days in American Firefighting*; Brendan McDonough's *My Lost Brothers: The Untold Story by the Yarnell Hill Fire's Lone Survivor*; and Kyle Dickman's *On the Burning Edge: A Fateful Fire and the Men Who Fought It*). We also relied on the study *Should I Stay or Should I Go Now? Or Should I Wait and See? Influences on Wildfire Evacuation Decisions* by Sarah McCaffrey, Robyn Wilson, and Avishek Konar.

Here and throughout the book our understanding of wildfires was rooted in interviews with firefighters and textbooks such as *Wildlands Firefighting Fundamentals* by William C. Teie and Brian F. Weatherford.

Chapter 2: Off the Grid

For information on PG&E, we relied on company documents; public records requests with the California Public Utilities Commission; court filings; the California Independent System Operator transmission plan; the NorthStar Consulting Report; the National Transportation Safety Board report on the San Bruno explosion; Charles M. Coleman's *P.G. and E. of California: The Centennial Story of Pacific Gas and Electric Company, 1852–1952*; and *The Smartest Guys in the Room: The Amazing Rise and Scandalous Fall of Enron* by Bethany McLean and Peter Elkind. We also spoke with linemen who wished to remain anonymous and representatives of the International Brotherhood of Electrical Workers Local 1245, including Tom Dalzell, Robert L. Dean Jr., and Ralph M. Armstrong. We drew details on the age of the transmission tower implicated in the Camp Fire from a PG&E legal filing prompted by a report in the *Wall Street Journal* as well as from the *New York Times*. California's energy history was gleaned from James C. Williams's *Energy and the Making of Modern California*, while the theory of California as a "hydraulic society" is elucidated by Donald Worster in *Under Western Skies: Nature and History in the American West*. Information on PG&E's Stairway of Power came from the National Oceanic and Atmospheric Administration; Deer Creek Resources; and Western Pacific Railroad History Online. PG&E did not respond to repeated requests for comment.

Details on the fires and explosions mentioned in this chapter came from Cal Fire reports; archives of the *Paradise Post*; the 2008–2009 Butte County Grand Jury Report; the study *Historical patterns of wildfire ignition sources in California ecosystems* by Jon E. Keeley and Alexandra D. Syphard; and reporting by the *San Francisco Chronicle*, SFGate.com, the Associated Press, the *Union*, the *Calaveras Enterprise*, the *Sacramento Bee*, the *Los Angeles Times*, the *Santa Rosa Press Democrat*, the *Mercury News*, E&E News; and the *Guardian*. Details about Butte County's response to the recommendations made in the 2008–2009 grand jury report came from a letter sent to the body from the Board of Supervisors.

The *Paradise Post* provided insight on life in the town leading up to the fire. Information on the Carr Fire came from Cal Fire, the *Redding Record-Searchlight*, the *San Francisco Chronicle*, the Associated Press, the *Guardian,* and the Facebook group Paradise Rants and Raves. Footage of a town council voter's forum came from the Rotary Club of Paradise and League of Women Voters town council candidate forum.

Chapter 3: Firebrands

Documents obtained from Cal Fire and Butte County, including the incident chronology report and transcripts and audio of 911 calls, helped us understand how the fire developed, as did interviews with Cal Fire commanders and firefighters, including Ken Pimlott, Scott Upton, Jim Messina, and others cited in the text. Upton was the commander who texted Pimlott when the fire broke out and later raced to Paradise. For the history of Cal Fire and information on its expenditures, we reviewed documents from the agency and the state, as well as reporting by the *Union Democrat* and *Air and Space Magazine*. Researcher Bob Yokelson explained the science of convection columns to us.

For Concow history, we interviewed the Moaks, Fred and Sally Hugg, Tony Salzarulo, Jacquelyn R. Chase, and others, and we found reporting by the *Chico Enterprise-Record* and the *Chico News & Review* helpful. Karen Williamson and Stanley Steven Jansson were the couple who left their dog in the car at Concow Reservoir. A fire tornado was also seen by Concow resident Stephanie Row. Sharon Beabout shared memories of Crystal Dave.

Chapter 4: Daybreak

The accounts detailed here were largely based on interviews with the people cited in the text. The unnamed individuals who saw portents of the fire are Jeremy Bredow; Joseph Kupstas, Don Willard, Sheila Craft, Krystin Harvey and her daughters Arissa and Arianne, and Marissa Nypl. The description of Rob Nichols's conversation with the dispatcher is based on Nichols's recollection. The timeline of Paradise's evacuation orders came from the Butte County Sheriff's Office. Transcripts of 911 calls were from Cal Fire, the Butte County Sheriff's Office, and the Paradise Police Department. Data on the number of people who signed up for the town's emergency-warning system and for the number who received warnings were from the Butte County Sheriff's Office and the Butte County 2018–2019 Grand Jury Report. Some details on evacuation alerts came from the *Los Angeles Times*. The woman who received an alert to evacuate as she watched her home burn was Anastasia Skinner. The high-ranking officials who directed traffic included the Paradise police chief Eric Reinbold, the Butte County sheriff Kory Honea, and the town councilman Mike Zuccolillo.

Some details about the hospital evacuation came from the *NFPA Journal*. Information on hospital staff and patients trapped in a home after an ambulance broke down came from stories in the *Washington Post, Los Angeles Times,* and Fox 40. Body-camera footage from deputy Aaron Parmley, provided by the Butte County Sheriff's Office, showed his journey through the burning town. Data on the scale of the Cal Fire response was provided by the organization. The Reno, Nevada, incident meteorologist is Alex Hoon. The details about the size of the fire and its rate of spread came from Cal Fire's chronology of events. Deer Creek Resources provided images that showed the fire's path through town.

Chapter 5: Stay or Go

Faith Kearns, Lincoln Bramwell, Stephen J. Pyne, Chris Dicus, and Alistair Smith all discussed the paradigm shift from wildfires to urban fires in interviews. The Jerry Williams quote is from Michael Kodas's *Megafire: The Race to Extinguish A Deadly Epidemic of Flame*. As in previous chapters, details about the timing of the fire's path, including reports from firefighters that it was burning in Butte Creek Canyon, came from Cal Fire's chronology of events. The reconstruction of Andrew Downer's final hours is based on interviews with Iris Natividad and telephone records. The reconstruction of John Sedwick's final hours is based on interviews with Skye Sedwick, Simona MacAngus, Waylon Shipman, Lloyd Romine, and Ray Johnson, as well as a voicemail Sedwick left that night.

Chapter 6: The Cemetery

Gene Woodcox's story was based on interviews with Matt Strausbaugh, a relative who posted the video on YouTube. John Sedwick's historical writings were published in *Tales of the Paradise Ridge*. The story of the Mann Gulch fire is best told in Norman Maclean's *Young Men and Fire*.

Chapter 8: A City Dispersed

This chapter was based largely on our original reporting on the fire, including press conferences and community meetings we attended and evacuation centers we visited, as well as ridealongs with the Paradise police. Reporting from KQED, the *San Francisco Chronicle*, the *Guardian*, the *Chico Enterprise-Record*, LA Taco, SF Gate, and Scriptype Publishing also aided us at various points in this chapter, particularly stories about air quality, donations, and missing Paradise residents. The Bay Area Air Quality Management District provided information about the path of smoke from the fire and details about inversion layers. For information on smoke waves we relied on the studies *Particulate Air Pollution from Wildfires in the Western US under Climate Change* by Jia Coco Liu and colleagues; and *Future Fire Impacts on Smoke Concentrations, Visibility, and Health in the Contiguous United States* by B. Ford and colleagues. Estimates of the number of people who evacuated, and how many ended up in shelters, came from Butte

County and the Butte County Sheriff's Office. Scott McLean was the Cal Fire spokesman quoted saying, "The town is devastated, everything is destroyed," in an interview with Reuters. Details about the number of evacuees in shelters who caught norovirus came from the Butte County Public Health Department.

Chapter 9: Search and Recovery

Harold Schapelhouman of the Menlo Park Fire District coordinated access to the team that responded to the Camp Fire. Other details about the identification of remains came from the *San Francisco Chronicle*. Accounts of life behind the cordon were provided by Jeff Evans, Pete Moak, Fred Hugg, Scott Carlin, and others. The victims of the "lottors" were Tamra South and her family. A recording of the meeting of the Paradise town council was made available by the town. The recounting of Donald's Trump visit and Thanksgiving were based on our reporting.

Chapter 10: A Pile of Ashes

Reports from the *Chico Enterprise-Record* and KRCR were useful in describing the city's reopening, as were our own visits to the area. The college student searching for embarrassing items in the rubble did not wish to be identified. Details about water contamination in Paradise came from information provided by the Paradise Irrigation District and reporting from the *Chico Enterprise-Record*, the *Chico News and Review*, and the *Sacramento Bee*. Figures on the number of homes that sold in Chico immediately after the fire came from reporting by the *Chico Enterprise-Record*. Statistics about Butte County's vacancy rate came from the county housing authority. Information about the county's capacity to house evacuees and related quotes from housing director Ed Mayer came from the *Sacramento Bee*.

Elderly people who died soon after the fire included, in addition to Barbara Allen, Nancy Clements and Patricia Anthony. We also relied on reporting from the *Chico News and Review*.

The quote from the school district superintendent came from a community meeting that we attended. Footage of Chelsea West's meeting with Bernie Sanders was published by the senator's office, and Allyn Pierce's appearance in the California Capitol was streamed by ABC10. White Pony Express was the nonprofit group that distributed food to displaced residents at FEMA sites.

Chapter 11: The Perfect Fire

Information about PG&E after the Camp Fire came from documents filed with the CPUC, court transcripts, lawsuits, as well as reporting by the *Wall Street Journal*, the *San Francisco Chronicle*, and Bloomberg. Details about Judge William Alsup came from a profile by the American Bar Association as well as reporting by The Verge. Details on the visit of company executives to Paradise came from the *San Francisco Chronicle*.

Information on the winner of the Donkey Derby came from the *Chico Enterprise-Record*. The study of gentrification after Hurricane Katrina is *Gentrification in the wake of a hurricane: New Orleans after Katrina* by Eric Joseph van Holm and Christopher K. Wyczalkowski. The Paradise population forecast was published by the *Chico Enterprise-Record*.

Epilogue

Details about the Kincade Fire came from Cal Fire. Information about the PG&E power shutoffs came from company announcements and reporting by the *Guardian*.